Mircea Gh. Negoita and Sorin Hintea

Bio-Inspired Technologies for the Hardware of Adaptive Systems

Studies in Computational Intelligence, Volume 179

Editor-in-Chief
Prof. Janusz Kacprzyk
Systems Research Institute
Polish Academy of Sciences
ul. Newelska 6
01-447 Warsaw
Poland
E-mail: kacprzyk@ibspan.waw.pl

Further volumes of this series can be found on our homepage: springer.com

Vol. 158. Marina Gavrilova (Ed.)
Generalized Voronoi Diagram: A Geometry-Based Approach to Computational Intelligence, 2009
ISBN 978-3-540-85125-7

Vol. 159. Dimitri Plemenos and Georgios Miaoulis (Eds.)
Artificial Intelligence Techniques for Computer Graphics, 2009
ISBN 978-3-540-85127-1

Vol. 160. P. Rajasekaran and Vasantha Kalyani David
Pattern Recognition using Neural and Functional Networks, 2009
ISBN 978-3-540-85129-5

Vol. 161. Francisco Baptista Pereira and Jorge Tavares (Eds.)
Bio-inspired Algorithms for the Vehicle Routing Problem, 2009
ISBN 978-3-540-85151-6

Vol. 162. Costin Badica, Giuseppe Mangioni, Vincenza Carchiolo and Dumitru Dan Burdescu (Eds.)
Intelligent Distributed Computing, Systems and Applications, 2008
ISBN 978-3-540-85256-8

Vol. 163. Pawel Delimata, Mikhail Ju. Moshkov, Andrzej Skowron and Zbigniew Suraj
Inhibitory Rules in Data Analysis, 2009
ISBN 978-3-540-85637-5

Vol. 164. Nadia Nedjah, Luiza de Macedo Mourelle, Janusz Kacprzyk, Felipe M.G. França and Alberto Ferreira de Souza (Eds.)
Intelligent Text Categorization and Clustering, 2009
ISBN 978-3-540-85643-6

Vol. 165. Djamel A. Zighed, Shusaku Tsumoto, Zbigniew W. Ras and Hakim Hacid (Eds.)
Mining Complex Data, 2009
ISBN 978-3-540-88066-0

Vol. 166. Constantinos Koutsojannis and Spiros Sirmakessis (Eds.)
Tools and Applications with Artificial Intelligence, 2009
ISBN 978-3-540-88068-4

Vol. 167. Ngoc Thanh Nguyen and Lakhmi C. Jain (Eds.)
Intelligent Agents in the Evolution of Web and Applications, 2009
ISBN 978-3-540-88070-7

Vol. 168. Andreas Tolk and Lakhmi C. Jain (Eds.)
Complex Systems in Knowledge-based Environments: Theory, Models and Applications, 2009
ISBN 978-3-540-88074-5

Vol. 169. Nadia Nedjah, Luiza de Macedo Mourelle and Janusz Kacprzyk (Eds.)
Innovative Applications in Data Mining, 2009
ISBN 978-3-540-88044-8

Vol. 170. Lakhmi C. Jain and Ngoc Thanh Nguyen (Eds.)
Knowledge Processing and Decision Making in Agent-Based Systems, 2009
ISBN 978-3-540-88048-6

Vol. 171. Chi-Keong Goh, Yew-Soon Ong and Kay Chen Tan (Eds.)
Multi-Objective Memetic Algorithms, 2009
ISBN 978-3-540-88050-9

Vol. 172. I-Hsien Ting and Hui-Ju Wu (Eds.)
Web Mining Applications in E-Commerce and E-Services, 2009
ISBN 978-3-540-88080-6

Vol. 173. Tobias Grosche
Computational Intelligence in Integrated Airline Scheduling, 2009
ISBN 978-3-540-89886-3

Vol. 174. Ajith Abraham, Rafael Falcón and Rafael Bello (Eds.)
Rough Set Theory: A True Landmark in Data Analysis, 2009
ISBN 978-3-540-89886-3

Vol. 175. Godfrey C. Onwubolu and Donald Davendra (Eds.)
Differential Evolution: A Handbook for Global Permutation-Based Combinatorial Optimization, 2009
ISBN 978-3-540-92150-9

Vol. 176. Beniamino Murgante, Giuseppe Borruso and Alessandra Lapucci (Eds.)
Geocomputation and Urban Planning, 2009
ISBN 978-3-540-89929-7

Vol. 177. Dikai Liu, Lingfeng Wang and Kay Chen Tan (Eds.)
Design and Control of Intelligent Robotic Systems, 2009
ISBN 978-3-540-89932-7

Vol. 178. Swagatam Das, Ajith Abraham and Amit Konar
Metaheuristic Clustering, 2009
ISBN 978-3-540-92172-1

Vol. 179. Mircea Gh. Negoita and Sorin Hintea
Bio-Inspired Technologies for the Hardware of Adaptive Systems, 2009
ISBN 978-3-540-76994-1

Mircea Gh. Negoita
Sorin Hintea

Bio-Inspired Technologies for the Hardware of Adaptive Systems

Real-World Implementations and Applications

Springer

Prof. Mircea Gh. Negoita
24 Fiona Grove
Karori
Wellington 6012
New Zealand
E-mail: mnegoita@hotmail.com

Prof. Dr. ing. Sorin Hintea
Technical University of Cluj-Napoca
Faculty of Electronics
Telecommunications and IT Basis of Electronics Department
str G. Baritiu nr 26-28
Cluj-Napoca 400027
Romania
E-mail: Sorin.Hintea@bel.utcluj.ro

ISBN 978-3-540-76994-1 e-ISBN 978-3-540-76995-8

DOI 10.1007/978-3-540-76995-8

Studies in Computational Intelligence ISSN 1860949X

Library of Congress Control Number: 2008942067

Typeset & Cover Design: Scientific Publishing Services Pvt. Ltd., Chennai, India.

Printed in acid-free paper

9 8 7 6 5 4 3 2 1

springer.com

To my country Australia – a world leader and promoter of high technology

To my native country Romania– an EU member of the world elite in high technology

Mircea Negoita

Foreword

Bio-Inspired Computing Technologies led to the spectacular progress of the Computational Intelligence (CI) nowadays and to its implementations in form of the Hybrid Intelligent Systems (HIS). Evolvable Hardware (EHW) has emerged as a novel and highly diversified technology and paradigm supporting the design, analysis and deployment of the high performance intelligent systems. The intellectual landscape of EHW is enormously rich. The discipline of EHW brings together hardware implementation of the main technologies of CI including fuzzy sets, neural networks, and evolutionary optimisation. But EHW systems use more than just the three broad areas mentioned above. They also cover novel areas as Artificial Immune Systems and DNA computing. The strength of EHW hinges on the synergy between these technologies supported by the advanced analogue and digital programmable circuits. This synergy helps exploit the advantages of the contributing technologies while reducing their possible limitations. The advanced programmable circuits confer the suitable hardware environment for a CI implementation from day to day more close to the intelligence of a human being.

EHW is a special case of the adaptive hardware, namely being strongly related to the Adaptive Systems (AS) and the Adaptive Hardware (AH).The progress in EHW is rapid. The individual technologies evolve quite quickly paving a way to new interesting and truly amazing applications. In the heart of all of those is the principle of hybridisation. EHW is suitable for the dramatic changes that happen in the relation between hardware and the application environment .This is in the case of malicious fault/ defects and need for new emergent functions that claim for in-situ synthesis of a totally new hardware configuration. It is not surprising at all witnessing a lot of activities and achievements within this realm included on the agenda of high tech organizations as NASA or ESA.

It is my pleasure indeed to introduce an attempt to provide the basis for EHW /AH through a comprehensive presentation of the fundamentals, key methods and the recent trends. The authors are two well-known researchers – both of them having a large industrial and academic experience: Professor Mircea Gh. Negoita of BLA Ltd, Brisbane, Australia and Professor Sorin Hintea of The Technical University of Cluj-Napoca, Romania. Both of them have a successful track record in this area and are highly qualified for this job. Consequently, the book has immensely benefited from their individual research, professional insights and practical experience.

I am convinced that the book will be a useful tool in pursuing further developments and applications of EHW/AH methods and circuits. The editors are to be congratulated on the careful selection of the material that very well reflects the breadth of the discipline covering a range of highly relevant and practical design principles governing the development of EHW/AH. Given this amazingly wide scope of the area, the reader will be pleased by the depth and clarity of exposure of the material. A newcomer will be pleased by the comprehensive and well-organized material of the volume. A practitioner will gain a down-to-the earth view at the design principles useful in the design, implementation, and validation of EHW/AH that rely on intelligent systems. This volume supplements very well the previous book by Negoita et al. entitled "Computational Intelligence. Engineering of Hybrid Systems" being already published in 2004 and the co-edited book by Negoita and Reusch entitled "Real World Applications of Computational Intelligence" already published in 2005.

The authors deserve our congratulations on their outstanding work. The authoritative coverage of the area is performed through a clear and well-organised way of presenting the fundamentals of key methods. This feature makes the book highly attractive. It is an excellent reading for everybody who is practically interested in the design and analysis of EHW/AH.

Charles C Nguyen, D.Sc.
Dean and Professor
Nguyen@cua.edu
October 15, 2008

School of Engineering
Catholic University of America
Washington DC, USA

Preface

Problems in engineering, computational science and the physical and biological sciences are using the increasingly sophisticated methods of Computational Intelligence (CI). This science/engineering field is mainly the result of an increasing merger of the stand alone Intelligent Technologies (IT), namely Fuzzy Systems (FS), Neural Networks (NN) , Evolutionary Computation (EC) , Artificial Immune Systems (AIS), DNA Computing and Knowledge Based Expert Systems (KBES). Because of the high interdisciplinary requirements featuring most real-world applications, no bridge exists between the different stand alone ITs. These technologies are providing *increasing benefit to business and industry*. The concomitant increase in dialogue and interconnection between the ITs has led to CI and its practical engineering implementation – the Hybrid Intelligent Systems (HIS).

Hardware implementation synergy of the main CI technologies including fuzzy sets, neural networks, and evolutionary optimization led to Evolvable Hardware (EHW). This is a novel and high performance technology and paradigm supporting the design, analysis and deployment of the high performance intelligent systems. EHW is a technical component typically featuring the most advanced structures of the Adaptive Systems (AS) and the most flexible Adaptive Hardware (AH). Nevertheless, it is hardware implementation of the most benefit for the society and indeed most revolutionizing application of CI by leading to the so-called EHW. These new CI based methodologies make possible the hardware implementation of both genetic encoding and artificial evolution, having a new brand of machines as a result. This type of machines is evolved to attain a desired behaviour that means they have a *behavioural computational intelligence*. There is no more difference between adaptation and design concerning these machines, these two concepts representing no longer opposite concepts. A dream of technology far years ago currently became reality: adaptation transfer from software to hardware is possible by the end. Much more, the electronics engineering as a profession was radically changed: the most soldering-based assembling manufacturing technologies are largely replaced now by technologies that use the strong technological support of advanced VLSI programmable circuitry, including EHW technologies. EHW can overcome a lot of technological manufacturing problems of the electronics integrated circuits: fabrication mismatches, drifts, temperature and other plagues to analog, exploiting the actual on-chip resources – finding a new circuit solution to the requirements with given constrains and actual on-chip resources.

EHW design methodology for the electronic circuits and systems is not a fashion. It is suitable for solving the special uncertain, imprecise or incomplete defined real-world problems, claiming a continuous adaptation and evolution too. Dramatic changes happen in the relation between hardware and the application environment, and this in case of *malicious faults* or need for *emergent new functions* that claim for in-situ synthesis of a totally new hardware configuration. EHW is suitable for flexibility and survivability of autonomous intelligent systems. EHW *survivability* means to maintain functionality coping with changes in hardware characteristics under the circumstances of adverse environmental conditions as for example: temperature variations, radiation impacts, aging and malfunctions. EHW *flexibility* means the availability to create new functionality required by changes in requirements or environment. The application developer may meet different design tasks to be evolved. As the case, the design to be evolved could be: a program, a model of hardware or the hardware itself. Algorithms that run outside the reconfigurable hardware, mainly feature the actual EHW state of the art. But also some chip level attempts were done. The path from chromosome to behaviour data-file is different in case of intrinsic and extrinsic EHW.

The progress in EHW is rapid. The expanding of EHW area of application nowadays is similar to Fuzzy technology in the last twenty years. Beginning with space and defense applications a few years ago, EHW/AHS is applied in humanoid robots for intelligent handling, EMG-prosthetic hand, and data compression for graphic arts, cellular phones, polymorphic electronics, self-repairing hardware and so on. It is not surprising at all witnessing the growing up of the EHW/AHS community as a distinct elite group inside the international scientific community of CI. The main EHW/AHS international conferences, workshops, symposia are supported by famous international research organizations, as NASA from USA or ESA from EU. Some reference research groups marked the previous and current trends and achievements of the EHW/AHS community at a world level: NASA JPL Evolvable Hardware Laboratory, USA; EHW Group at the Advanced Semiconductor Research Center of National Institute of Advanced Industrial Science and Technology (AIST), Japan; EHW Group at the Intelligent Systems Research Group of the University of York, UK; EHW Group at Department of Computer Systems, The Faculty of Information Technology (FIT), Brno University of Technology, Czech Republic; The Reconfigurable and Embedded Digital Systems Institute (REDS) at The University of Applied Sciences (HEIG-VD), Switzerland.

The idea of writing a book on this topic first crossed my mind in 2001, and I am really happy that the book is finally complete. Initially I thought this book would be of real help to my gifted students at the School of IT at Wellington Institute of Technology, Wellington, New Zealand. New ideas and suggestions crucially guiding the final structure of the book were used as a result of my research visit at NASA JPL in 2003 and as a result of my direct contacts and involvement in organizing the NASA/DoD EHW series of conferences.

The purpose of this book is to illustrate the current needs and to emphasize the future needs for the interaction between various component parts of the EHW/AHS framework. The team writing this book did this firstly by encouraging the ways that EHW techniques may be applied in those areas where they are already traditional, as well as pointing towards new and innovative areas of application involving emergent technologies such as Artificial Immune Systems. Secondly, the team aimed to help encourage other disciplines to engage in a dialogue with practitioners of EHW/AHS engineering, outlining their problems in accessing these new methods in the engineering of intelligent AS, and also suggesting innovative developments within the area itself. Thus the progress of EHW/AH within the framework of intelligent AS was discussed from an application - engineering point of view, rather than from a cognitive science or philosophic view point. In this respect, regarding the technological support of EHW/AH the team of authors is most focused on the analog reconfigurable hardware that has actually a huge weight in the EHW environment. The appearance of reconfigurable analogue arrays (FPAAs) was crucial for the technological support required by companies involved in electronics research and development as well as in manufacturing. The analog reconfigurable hardware allows prevention or removal of essential fabrication mismatches and other refined technological problems by evolving circuits as the case. Practical engineering comments are made regarding the so-called *custom made* EHW-*oriented reconfigurable hardware* that can reprogram many times, can understand what's inside and is featured by a flexible programmability.

Beside some concrete elements of the EHW design, the book delivers a global image of the current limits in evolutionary design of AH. Also the practitioners get an accurate image of two different and distinct approaches governing EHW/AH: *Evolutionary circuit design* performs the evolution (the design) of a single circuit, with additional features such as fault tolerance, testability, polymorphic behaviour, that are difficult to design by conventional methods; *Evolvable hardware* involves an evolution responsible for continual adaptation applied to high-performance and adaptive systems in which the problem specification is unknown beforehand and can vary in time

Due to the best efforts of both co-authors, the book looks like a homogenous work aimed to be accessed very comfortable to a large range of public.

Chapter 1 is an introduction to Computational Intelligence and Intelligent Hybrid Systems, focused on their terminology and classification connections to EHW/AH. Emergent Intelligent Technologies implication on the Adaptive Hardware Systems is presented. A special part describes the AIS as a special technology for the Adaptive Systems.

Chapter 2 presents an engineering perspective of the EHW terminology, design methods and application relied on very practical engineering remarks. Direct application related aspects of immediate help for any EHW practitioners are another topic of this chapter. Some EHW specific programmable integrated circuits are introduced, especially the new generation – field programmable gate arrays (FPGAs) and most recently reconfigurable analogue arrays (FPAAs) and field-programmable interconnection circuits (FPICs) or configurable digital chips at the functional block level, (open-architecture FPGAs).

Chapter 3 presents a number of Genetic Algorithms applications in the field of electronic circuits design, both analog and digital. Examples of logic circuits are introduced, including arithmetic circuits designed using evolutionary techniques, and the results are compared with the conventional methods. Another promising field of application is consisting of reconfigurable circuits applied in mobile communications, as a part of more complex adaptive systems. Also, there are proposed biomedical applications such as the implanted auditory prosthesis and other electronic stimulators. It is included a description of an AO design method based on fuzzy techniques and genetic algorithms.

Most sub-chapters include useful suggestions for the practical design and development of further applications. Both authors agree that although this book is a primer, it is not useful to only students. This book has practical value for both those new to the discipline and also for those who are already practitioners in the area.

The common research work and exchange of ideas with my distinguished colleague – constitutes the foundation for this book. Professor Sorin Hintea, from the Technical University of Cluj-Napoca, from my native country - Romania is a world known personality acting inside a very promising EHW/AH high tech European research group.

A decisive element for finally completing the book was the support of EHW/AH Group at NASA JPL, the special remarks and advice from its leader - dr. Adrian Stoica. Special thanks are due to Prof. Lukas Sekanina , Faculty of Information Technology, Brno University of Technology, for our permanent interactive scientific connection.

The authors are grateful for the understanding and permanent support of Springer Verlag Publishing House throughout the writing of this book. We would also like to acknowledge our special appreciation for the permanent support of dr. Robert J. Howlett - the Executive Chairman of KES International Organization - a leading professional organization that strongly supports and promotes EHW/AH technologies, conferences and publications.

On behalf of both authors,

Queensland, Australia
October 15, 2008

Bongaree, Bribie Island
Prof. Mircea Gh. Negoita

Contents

1 Bio-Inspired Computational Intelligence for the Hardware of Adaptive Systems.. 1

1.1 Techniques of Computational Intelligence........................... 1

1.2 Features and Classifications of Hybrid Intelligent Systems......... 9

1.3 Emergent Intelligent Technologies and the Adaptive Hardware Systems: AIS – A Technology for the Adaptive Systems.......... 13

2 Advanced Hardware Implementation of the Computational Intelligence and Intelligent Technologies.. 27

2.1 Evolvable Hardware: An Overview.. 28

2.1.1 EHW Classification, Practical Engineering Remarks......... 30

2.1.2 EHW Technological Support (FPGA, FPAA, FPTA, FPMA, PsoC).. 37

2.1.2.1 Introduction to Programmable Integrated Circuits... 38

2.1.2.2 FPGA Families and Advanced Type of FPGA....... 40

2.1.2.3 Field Programmable Analog Arrays(FPAA)......... 44

2.1.2.4 Field Programmable Transistor Arrays (FPTA) Produced by NASA JPL.............................. 46

2.1.2.5 Practical Remarks on the Technological Support of EHW.. 48

2.1.3 EC Based Methods in EHW Implementation: EHW Architectures... 50

2.1.4 An Application of GA for the Design of EHW Architectures.. 63

2.1.5 Global Remarks on Current Methods in EHW Technology and Its Prospectus... 70

2.2 Hardware Implementation of the Artificial Immune Systems...... 73

2.3 Hardware Implementation of DNA Computing...................... 75

2.4 Elements of Intercommunications Inside the AHS/EHW International Community (Conferences; Books; Journals; Elite Departments)... 76

3 Bio-Inspired Analogue and Digital Circuits and Their Applications... 83

3.1 Introduction.. 83

3.2 Genetic Algorithms for Analogue Circuits Design.................. 84
3.2.1 GA as Tools to Design Analogue Circuits.................... 84
3.2.2 Overview of the Genetic Algorithm............................ 86
3.2.3 Representation.. 89
3.2.4 Analogue Applications with FPTA Cells..................... 91
3.2.5 Design Optimization of a CMOS Amplifier.................. 92
3.2.5.1 Formulation of the Optimization Problem.......... 92
3.2.5.2 Evaluation Engine....................................... 94
3.2.5.3 Optimization Engine.................................... 94
3.2.5.4 Design Optimization of a CMOS Amplifier........ 95
3.2.6 Evolving Software Models of Analogue Circuits............ 97
3.3 Evolutionary Design of Digital Circuits............................... 99
3.3.1 Combinational Logic Circuits Evolutionary Design......... 99
3.3.2 Conventional Design Techniques for Arithmetic Adders and Multipliers... 102
3.3.2.1 One Bit Full Adders..................................... 102
3.3.2.2 Parallel Processing Adder............................. 105
3.3.2.3 CMOS Gates Full Adder................................ 106
3.3.2.4 The Mirror Adder.. 107
3.3.2.5 Full Adder with CMOS Transmission Gates....... 107
3.3.2.6 Serial Processing Adder................................ 109
3.3.2.7 Conventional Binary Multipliers..................... 109
3.3.3 Arithmetic Circuits Designed with Evolutionary Algorithms.. 112
3.3.3.1 Full Adders Design..................................... 112
3.3.3.2 Gate-Level Evolutionary Design..................... 114
3.3.3.3 Binary Multipliers Designed with Evolutionary Algorithms.. 116
3.3.4 Concluding Remarks on Digital Circuits Evolutionary Design.. 118
3.4 Reconfigurable Analogue Circuits in Mobile Communications Systems... 119
3.4.1 Multi-standard Terminals for Mobile Telecommunications.. 119
3.4.2 Reconfigurable Multi-Standard Analogue Baseband Front-End Circuits in Mobile Communications Systems... 122
3.4.3 Reconfigurable RF Receiver Architectures.................. 125
3.4.3.1 Superheterodyne Receiver........................... 125
3.4.3.2 Direct - Conversion Architecture................... 126
3.4.3.3 Low IF Architecture.................................... 127
3.4.3.4 Software Defined Radio............................... 127
3.4.3.5 Digital – IF Receiver................................... 129
3.4.4 Fully Reconfigurable Analogue Filters Design............ 129
3.4.5 Reconfigurable Filter Stage for a Combined Zero-IF/Low-IF Radio Architecture........................ 131
3.4.5.1 Flexible Zero-IF/Low-IF Radio Architecture..... 131

3.4.5.2 Transconductor-Based Reconfigurable and Programmable Analogue Array 133
3.4.5.3 Modular G_m-C State-Variable "Leapfrog" Filters.. 135
3.4.5.4 Simulation Results.................................... 139
3.4.5.5 Conclusions... 140
3.4.6 Variable Gain Amplifiers.. 141
3.4.7 Genetic Algorithms for Reconfigurable Analogue IF Filters Design.. 146
3.5 Biomedical Engineering Applications................................ 148
3.5.1 Electrical Stimulation and Neural Prosthesis............... 148
3.5.2 Cochlear Prosthesis via Telemetric Link..................... 150
3.5.3 Reconfigurable Circuits in Implantable Auditory Prosthesis.. 151
3.5.4 AGCs in Auditory Prosthesis................................. 153
3.5.5 Binary Controlled Variable Gain Amplifiers................ 156
3.5.5.1 Introduction.. 156
3.5.5.2 Digitally Controlled Gain Amplifier with Current Mirrors...................................... 156
3.5.5.3 Current Division Network.......................... 160
3.5.5.4 Programmable Amplifiers with CDN............. 162
3.5.5.5 Digitally Controlled Current Attenuator........... 165
3.6 Concluding Remarks.. 168

References... 169

List of Acronymes

A	**A***denine (a* **DNA** *base)*
ABR	**A***rchitecture* **B***it* **R***egisters*
ADC	**A***nalogue – to -* **D***igital* **C***onverter*
AEH	**A***daptive and* **E***volvable* **H***ardware*
AGC	**A***utomatic* **G***ain* **C***ontrol*
AH	**A***daptive* **H***ardware*
AI	**A***rtificial* **I***ntelligence*
AIS	**A***daptive* **I***mmune* **S***ystem*
AIS	**A***rtificial* **I***mmune* **S***ystem*
AMR	**A***dvanced* **RISC M***achine,*
ANFIS	**A***daptive* **N***euro -* **F***uzzy* **I***nference* **S***ystem*
APC	**A***ntigen-***P***resenting* **C***ells*
AS	**A***daptive* **S***ystem*
ASIC	**A***pplication-***S***pecific* **I***ntegrated* **C***ircuit*
B-cells	**B***-lymphocytes*
BCA	**B-c***ell* **A***lgorithm*
BDD	**B***inary* **D***ecision* **D***iagram*
BDT	**B***inary* **D***ecision* **T***ree*
BioMEMS	**Bio***medical* **MEMS**
BIST	**B***uilt –* **I***n –* **S***elf –* **T***est*
BPF	**B***and* **P***ass* **F***ilter*
C	**C***ytosine*
CA	**C***ellular* **A***utomata*
CAP	**C***onfigurable* **A***nalog* **P***rocessor*
CDN	**C***urrent* **D***ivision* **N***etwork*
CI	**C***omputational* **I***ntelligence*
CPLD	**C***omplex* **P***rogrammable* **L***ogic* **D***evices*

CPU	*Central Processing Unit*
CWP	*Computing with Word and Perception*
DAC	*Digital – to - Analogue Converter*
DCR	*Direct – Conversion Receiver*
DGA	*Designer Genetic Algorithms*
DN	*Distribution Network*
DNA	*Deoxyribo Nucleic Acid*
DNA-AIS	*Hybridization between* **DNA** *systems and* **AIS**
DNA-FS	*Hybridization between* **DNA** *systems and* **FS**
DNA-NN	*Hybridization between* **DNA** *systems and* **NN**
DNA-GA	*Hybridization between* **DNA** *systems and* **GA**
DSP/ASIC	*Digital Signal Processor/* **ASIC** *(a hybrid solution onto a configurable core featured by the best of both* **DSP** *and* **ASIC** *architectures)*
EA	*Evolutionary Algorithms*
EC	*Evolutionary Computation*
EC -AIS	*Hybridization between* **EC** *systems and* **AIS**
EHW	*Evolvable HardWare*
EHW-AIS	*Hybridization between* **EHW** *systems and* **AIS**
ES	*Evolutionary Strategy*
EP	*Evolutionary Platform*
EUNITE	*European Network on Intelligent Technologies for*
EW	*Electronic Warfare*
FES	*Functional Electrical Stimulation*
FNS	*Functional Neuromuscular Stimulation*
FPAA	*Field Programmable Analog Arrays*
FPGA	*Field Programmable Gate Arrays*
FPIC	*Field Programmable Interconnection Circuits*
FL	*Fuzzy Logic*
FPMA	*Field Programmable Mixed-signal Array*
FPTA	*Field Programmable Transistor Arrays*
FPPA	*Field Programmable Processor Arrays*
FS	*Fuzzy System*
FS-AIS	*Hybridization between* **AIS** *systems and* **FS**
FSM	*Finite State Machines*
G	*Guanine*
GA	*Genetic Algorithm*

GA-FS	**H***ybridization between* **GA** *systems and* **FS**
GA-NN	**H***ybridization between* **GA** *systems and* **NN**
GA-NN-FS	**H***ybridization between* **GA** *systems,* **NN** *systems and* **FS** *systems*
GAP	**G***enetic* **A***lgorithm* **P***rocessor*
GP	**G***enetic* **P***rogramming*
HIS	**H***ybrid* **I***ntelligent* **S***ystems*
HOT	**H***ardware* **O***bject* **T***echnology*
HPRCs	**H***igh* **P***erformanc*e **R***econfigurable* **C***omputers*
HPF	**H***igh* **P***ass* **F***ilter*
I/O	**I**nput/**O***utput interface*
IIM	**I***nnate* **I***mmune* **S***ystem*
IT	**I***ntelligent* **T***echnologies*
ITS	**I***ntelligent* **T***utoring* **S***ystems*
IT-2 FP	**I***nterva*l **T***ype-***2 F***uzzy* **P***rocessors*
IT2-FLS	**I***nterva*l **T***ype-***2 F***uzzy* **L***ogic* **S***ystem*
JPL	(**NASA**) **J***et* **P***ropulsion* **L***aboratory*
KBES	**K***nowledge* **B***ased* **E***xpert* **S***ystems*
KDD	**K***nowledge* **D***iscovery in* **D***ata Bases*
LNA	**L***ow* - **N***oise* **A***mplifier*
LPF	**L***ow* **P***ass* **F***ilter*
MC	**M***olecular* **C***omputing*
MEMS	**M***icro***-E***lectro***-M***echanical* **S***ystems*
MPU	**M***ain* **P***rocessing* **U***nit*
MLCEA-TC	**M***ulti-***L***ayer* **C***hromosome* **E***volutionary* **A***lgorithm –* **T***ransistor* **C***ount*
NESW	**N***orth-* **E***ast-***S***outh-***W***est interconnection*
NGA	**N***agoya* **GA**
NN	**N***eural* **N***etworks*
NW-AIS	**n***etwork-***b***ased* **AIS**
OTA	**O***perational* **T***rans-conductance* **A***mplifiers*
PAL	**P***rogrammable* **A***rray* **L***ogic*
PAMA	**P***rogrammable* **A***nalog* **M***ultiplexer* **A***rray*
PB-AIS	**p***opulation-***b***ased* **AIS**
PCR	**P***olymerase* Chain **R***eaction*
PCI	**P***eripheral* **C***omponent* **I***nterconnect* Standard
PDC	**P***ersonal* **D***igital* **C***ellphone*

PFU	**P***rogrammable* **F***loating* **U***nit*
PGA	**P***rogrammable* **G***ain* **A***mplifier*
PLA	**P***rogrammable* **L***ogic* **A***rray*
PLD	**P***rogrammable* **L***ogic* **D***evices*
PTA	**P***rogrammable* **T***ransistor* **A***rray*
PROM	**P***rogrammable* **R***ead-***O***nly* **M***emory*
PSO	**P***article* **s***warm* **o***ptimization*
PSO-DR	**P***article* **s***warm* **o***ptimization with* **d***iscrete* **r***ecombination*
PSoC	**P***rogrammable* **S***ystem-***o***n-***C***hip*
QC	**Q***uantum* **C***omputing*
RAT	**R***adio* **A***ccess* **T***echnologies*
RH	**R***econfigurable* **H***ardware*
RHIS	**R***obust (***S***oft* **C***omputing)* **I***ntelligent* **S***ystems*
RISC	**R***educed* **I***nstruction* **S***et* **C***omputer architecture*
RLC	**R***einforcement* **L***earning* **C***omponent*
RLD	**R***econfigurable* **L***ogic* **D***evice*
RM	**R***econfigurable* **M***echanism* (**RM**)
RNA	**R***ibo* **N***ucleic* **A***cid*
RPU	**R***econfigurable* **P***rocessing* **U***nit*
SABLES	**S***tand-***A***lone* **B***oard-***L***evel* **E***volvable* **S***ystem*
SAS	**S***mart* **A***daptive* **S***ystem*
SC	**S***oft* **C***omputing*
SDR	**S***oftware* **D***efined* **R***adio*
SGA	**S***imple* **GA** *(A classical Goldberg's* **GA**)
SME	**S***mall and* **M***edium* **E***nterprises*
SoPC	**S***ystems* **o***n* **P***rogrammable* **C***hip*
SRAM	**S***tatic* **R***andom* **A***ccess* **M***emory*
SS	**S***mart* **S***ystem*
S-W	**S***mith-***W***aterman* algorithm
T-cells	**T***-lymphocytes (T-cells*
T	**T***hymine*
TI DSP	**T***exas* **I***nstrument***s** **DSP**
T2-FLS	**T***ype-***2** **F***uzzy* **L***ogic* **S***ystem*
TRAC	**T***otally* **R***ecofigurable* **A***nalog* **C***ircuit*
UMTS	**U***niversal* **M***obile* **T***elecommunications* **S***ystem*
VGA	**V***ariable* **G***ain* **A***mplifier*

VHDL	**V***ery* **H***igh-speed integrated Circuit* **H***ardware* **D***escription* **L***anguage*
VLGGA	**V***ariable* **L***ength* **G***enotype* **G***enetic* **A***lgorithm*
VLSI	**V***ery* **L***arge* **S***cale* **I***ntegration*
VPU	**V***ector* **P***rocessing* **U***nit*
WITNeSS	**W***ellington* **I***nstitute of* **T***echnology* **N***ovel* **E***xpert* **S***tudent* **S***upport*

Chapter 1
Bio-Inspired Computational Intelligence for the Hardware of Adaptive Systems

1.1 Techniques of Computational Intelligence

Computational **I**ntelligence (**CI**) is a soft computing framework with a large variety of efficient applications. Problems in engineering, computational science and the physical and biological sciences are using the increasingly sophisticated methods of **CI**. Because of the high interdisciplinary requirements featuring most real-world applications, no bridge exists between the different stand alone **I***ntelligent* **T***echnologies* (**IT**), namely **F***uzzy* **S***ystems* (**FS**), **N***eural* **N***etworks* (**NN**), **E***volutionary* **C***omputation (***EC**), **A***rtificial* **I***mmune* **S***ystems* (**AIS**), **DNA** *Computing* and **K***nowledge* **B***ased* **E***xpert* **S***ystems* (**KBES**). The **CI** field is mainly the result of an increasing merger of the stand alone **IT**s. These technologies are providing *increasing benefit to business and industry*. Most of the countries in the world have understood the important role of **CI**. Not only have governmental institutions support National Projects in this area, but a lot of private companies are also currently using intelligent technologies in a number of application areas. These applications are not the consequence of fashionable, trendy ideas but a response to real world needs (Negoita , Neagu and Palade 2005). The concomitant increase in dialogue and interconnection between the **IT**s has led to **CI** and its practical engineering implementation – the **H***ybrid* **I***ntellgent* **S***ystems* (**HIS**).

The **CI** has the power of *increasingly improve any area of our social-economic life*. Nowadays we are confronted with a lot of complex real-world applications. Examples would be different forms of pattern recognition (image, speech or handwriting), robotics, forecast and different kind of decision-making in uncertainty conditions, etc… These kind of applications rely on a new concept, a new framework that is named **S***oft* **C***omputing* (**SC**) by L.A. Zadeh, the father of fuzzy systems. The basic ideas underlying **SC** belong to Prof. Zadeh. His vision of different techniques of interaction is strongly influencing the development of the new generation of intelligent (perception-based) systems toward a **C**omputational **I**ntelligence (**CI**). New tools are developing for dealing with world knowledge, for example the **C***omputing* with **W***ords* and **P***erceptions* (**CWP**) - a **FS** based-method featured by the understanding that perceptions are described in a natural language (Zadeh and Nikravesh 2002), (Zadeh 2003).

M.G. Negoita, S. Hintea: Bio-Inspired Tech. for the Hardware of Adaptive Sys., SCI 179, pp. 1–25.
springerlink.com

It is of high importance for the practitioners to understand that **CI** and **A***rtificial* **I***ntelligence* (**AI**) are distinct frameworks and must be distinguished with the aim of improving their exploitation in applications. A useful comparison between the **CI** and **AI** frameworks was made in (Negoita , Neagu and Palade 2005), focussing on four main characteristics: the main purpose, the methodological framework, the nature of the information processing and the proven, successful applications to each of them.

Each **CI** and **AI** method has its particular *strengths* and weaknesses that make them suitable for some applications and not other. These specific limitations of **AI** and **CI** led to creation of **HIS**, where two or *more techniques are combined* and co-operate to overcome the limitations of each individual technique. An overview of the main individual **CI** and **AI** techniques will be useful with regard to their successful applications, the advantages and limitations manifesting the individual effectiveness of the method.

FS deals with the handling of imprecision in data, finally making a decision made by *fuzzy concepts* and *human like approximate reasoning*. **FS** have knowledge bases that, because they are based on fuzzy IF - THEN rules are easy to examine, understand, update and maintain. Another typical advantage of **FS** is their high degree of flexibility in dealing with incomplete and inconsistent data by aggregating the hypothesis of all the rules. But **FS** also has real limitations – the heavy *reliance on human experts* in building the fuzzy IF - THEN rules, the need for manual *specification* of membership functions and fuzzy rules, as well as strong restrictions in *automatic adaptive learning* capabilities in an ever-changing external environment.

NN deals with inherent *parallelism* in data flow. Their capacity to learn patterns in data that are noisy, incomplete and even contradictory, confer on **NN** two distinct advantages in processing information - the ability of *processing incomplete information* and *generalize the conclusions*. The same as **FS** has its limitations, so does **NN. NN**s are *unable to explain how they arrived at their conclusions* and are unable to interact with conventional databases. Also there is a *lack of structured knowledge representation*. Also **NN** have real *scalability problems* in that they experience great difficult in training and in generalization capabilities for large and complex problems.

Like **NN, EC** (**G***enetic* **A***lgorithms* - **GA**, especially) also deal with *parallelism* in data flow. **EC** provides efficient search and optimisation tools for very large data sets, largely unconstrained by local minima problems. They don't require any mathematical modelling. **EC** are optimisation methods for either symbolic or **NN** or **FS** systems. Strictly referring to **GA**, their effectiveness consists of a high suitability for parallel computer implementation; particular success in large search and optimisation problems; in an ability to learn complex relationships in incomplete data sets, their use as „data mining" tools for discovering previously unknown pattern; in their ability to provide an explanation of how decisions were produced, in a format that humans can understand; an ability to adapt to changes in their operating environment. **GA** limitations are as follows: they are *computational expensive* methods involving an *application dependent setting* of **GA** parameters (e.g. for crossover, mutation) that could be a time-consuming trial and error process.

DNA computing is certainly a promising implementation method of information-processing capabilities of organic molecules, used with the aim of replacing digital switching primitives in computers (Paun et al 1998). The technology is still struggling to combine electronic computing with **DNA** computing, where most of the DNA operation is still being carried out in test tubes without the intervention of the user. **DNA** computing is characterized by a massive parallelism conferred by **DNA** strands which means the first strength of **DNA** computing has much higher *parallel data structures* than either **NN** or **GA**. The second strength of **DNA** computing devices results from the *complementarity* between two **DNA** strands when bonding takes place. **DNA** effectiveness as a powerful tool for computing relies on the handling of suitable encoded information of a very high density that leads to far-reaching conclusions when bonding takes place. The main limitation of **DNA** computing consists of a *higher than admissible error rate of operations* due to an actual lack of a proper technological environment.

The idea of harnessing individual biomolecules operating at nanoscales for computational purposes started at the beginning of the last decade. Nowadays, self-assembly of **DNA**, **RNA** and protein molecules is a current theme in biomolecular computing not only for computer science purposes, but as an effective medium to enable the construction of objects capable of performing useful tasks in massive numbers at the nanoscales of basic constituents. Different kind of bio(biomolecular) machines are common these days: **DNA** tweezers, microfluidic reactors, molecular motors, megalibraries of noncrosshybridizing **DNA** oligos, **DNA** nanotubes and circuits, protein motif search rngines. These family of biomachines are aimed to compute under real physical and biochemical constraints, in real world application environmemt and not in simulated or sanitized environments (Garzon 2003). There are two major categories of work in biomolecular computing : “in vitro” machines and “in silico” simulations for design and analysis of their operation. The computation “in silico” is aimed to help the increasing of biomolecular computing in reliability, efficiency and scalability.

The reliability of a **DNA** protocol means a degree of confidence that the laboratory experiment provides a correct and true answer to a problem to be solved. The efficiency of a **DNA** protocol is regarding the intended and effective contribution of the molecules that intervene in it. The scalability of a **DNA** computation means the effective reproductibility of the experiment with longer molecules that can encode larger problem instances while still obtaining equally reliable results under comparable efficiency.

An accurate analysis of the efficiency and reliability of **DNA** computing protocols via high fidelity benchmarked simulations is made in (Garzon et al 2003). This research team is successful in illustrating the advantages of using **DNA**-like structure and algorithms to solve the problems even “ in silico”, as well as the advantages of this new type of sexual genomic representation for **EA**. This work is a serious starting point that opens the way of possibility that one dream of high technology might become reality. This is regarding the technology ability to build a device that enjoys both the versatility and thermodynamical efficiency afforded by biomolecules, as well as the high levels of reliability and efficiency that soli-state electronics have achieved in the last half of century.

There is a radical conceptual difference between **DNA** computing and the other biologically-inspired **IT** - the **AIS**, **NN** and **EC** mainly. This **IT** triplet develops a range of computational algorithms derived from ideas or metaphors from different biological systems, while **DNA** computing tries to use the **DNA** computation operators as a new computing paradigm. But the **DNA** strands as data structures are nothing else than a particular case of an **AIS** typical Symbolic shape-space represented by the quaternary alphabet {A, C, G, T}. A careful analysis was made on the operators in **DNA** computing (de Castro and Timms 2002). As a result, similar operations at the cellular level in the **AIS** were remarked. (Deaton at al 1997) proposed an implementation of the **AIS** negative selection algorithm in a **DNA**-based computer. Their idea was relied on viewing two single strand molecules by Watson-Crick complementarity as an **AIS** string matching. As a consequence, they performed the **DNA** implementation of negative selection algorithm by using techniques from molecular biology for performing the censoring and monitoring part of the **AIS** negative selection algorithm.

Table 1.1 The vocabulary of **AIS** mapping into **DNA** Computing

Terminology	**AIS**	**DNA Computing**
Encoding	Attribute strings in Shape-Space that represent immune cells and molecules	**DNA** strands composed of the nucleotides {A, C, G,T}
Processing	**AIS** algorithms and processes	**DNA** operations
Recombination	Mutation and gene rearrangement	Annealing
Binding	Shape complementarity in Shape-Space	Watson-Crick complementarity
Reproduction	Clonal expansion	Amplification via **P***olymerase* **C**hain **R***eaction* (**PCR**)
Antigen processing and presentation	Antigen fragmentation and presentation via **A***ntigen*-**P***resenting* **C***ells*) **APC**	Operations: melting, separate, detect and length-separate

The mapping of the **AIS** into **DNA** computing make definitely possible the interaction between these two **IT** (de Castro and Timms 2002), as in Table 1.1.

AIS most complete definition is as from (de Castro and Timmis 2002): << **AIS**s are adaptive systems, inspired by theoretical immunology and observed immune functions, principles and models, which are applied to problem solving >>

It would be absolutely wrong to limit us by thinking that **AIS** is just a methodology concerned only with the use of immune system components and processes as inspiration to construct computational systems. **AIS** means much more, namely it is a concept and framework that radically changes the design philosophy of any artificial systems and in the same time sensitively improves the system behaviour in the adverse external environment of application. This affirmation is made by relying on some considerations of reliability design that are deeply related to the drawbacks of the artificial systems towards error and faults and also are technically justified in connection with the concept of malicious defects.

All kind of actual artificial systems not reliant on **AIS** are limited in their behaviour regarding the wrong events, errors and faults namely. These systems are passive and event-driven because they usually wait for a specific wrong event and trigger a treatment specific to the event. Both the fault and its remedia can be apriori specified and enumerated and the system is dependent on the concept of redundancy: if a fault occurs that causes a loss of function, the function lost because of the faulty component is replaced by a redundant standby component (Ishida 2004).

AIS framework develops an attitude towards the errors and faults that is radically different; they are mainly relied on actively using the information from the application environment. They are ready to act and successfully annihilate any challenges posed by elements that are coming from a dynamic external environment involving a large range of unpredictable events. The high degree of complexity featuring the actual applications requirements and also the intrinsic systems architecture/functionality are difficult circumstances favouring an unexpected level of effect evolution of exogenous dangerous element on systems. These elements of real danger evolved so much that even can reproduce and spread.

The concept of malicious fault is a reality, see for example the malicious software that causes serious damages and troubles, even more worse that ordinary hardware or software faults are doing. The malicious defects must be viewed as the viruses, parasites and /or predators are for the biological systems.

Two structural details are crucial for both the terminology and comprehensibility of **AIS**:

- the element recognizing a particular pattern is called an *antibody* ;
- the pattern recognized by an antibody is called an *antigen*.

The affinity between an antibody and an antigen involves extensive regions of complementarity in the search space. This is the consequence of the condition in order for an antigen to be recognized by an antibody or another antigen: they must bind complementary with each other over an appreciable portion of their surfaces. The search space is called a Shape-Space, where each of its multidimensional points, specifies the generalized shape (dimension) of an antigen-binding region with regard to its antigen binding properties. The interaction between antibodies or of an antibody and an antigen is evaluated by the *affinity measure*. The affinity measure is evaluated by a distance measure between the corresponding attribute strings to the antibodies and antigens. The most like-biological affinity must take into account that the molecules are allowed to interact with each other in different alignments. As a conseqence, the calculated affinity must be a total affinity to be

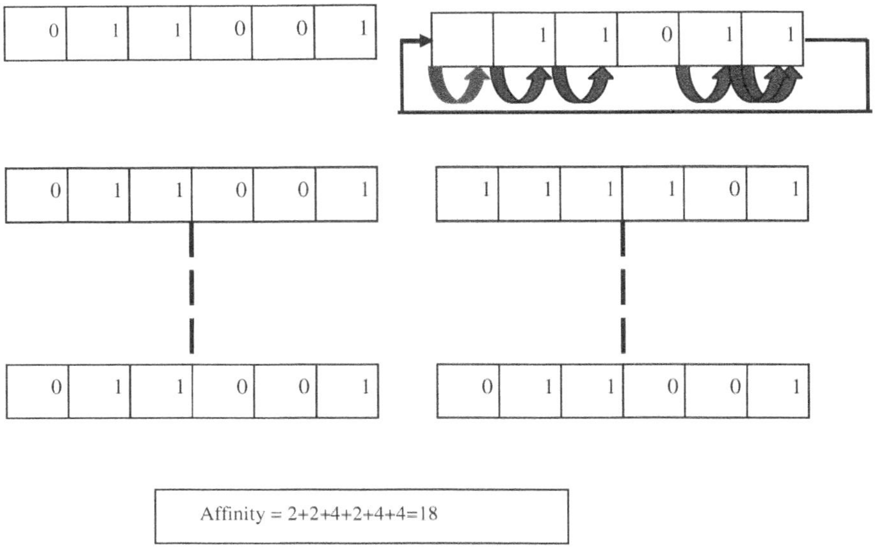

Fig. 1.1 Affinity of **AIS** bitstring-represented antibodies and antigens

calculated by summing the affinity featuring each possible molecular alignment. Fig 1.1 depicts a simplified schematic of this process for two binary strings of length 6: one string is fixed and its complementary string is left-to-right rotated. The process is repeated until the flipping string return to its initial configuration by crossing all possible alignments.

Two structural details are crucial for both the terminology and comprehensibility of an **AIS**: the element recognizing a particular pattern is called an *antibody;* the pattern recognized by an antibody is called an *antigen.*

As a conclusion, the role of **AIS** concept and framework is essential in developing a crucial strategy for of systems survival - the intelligent defence of the information systems against malicious faults. The artificial immunity-based systems are complete information systems featured by three main informational features (Ishida 1996), (Dasgupta et al 1996), (Negoita , Neagu and Palade 2005):

- the *self-maintenance* property involving monitoring not only of the nonself (antigens) but also of the self and this based on the self (antibodies) counterpart of the system
- the property of a *distributed system* structured by autonomous components having a capability of mutual evolution that results in forming an ad hoc network with specific recognition
- the property of *an adaptive system* featured by diversity and selection, based on selection as opposed to instruction

These three informational features define an interesting view for guiding the **AIS** development as strongly inspired by the natural Immune System, see Fig. 1.2.

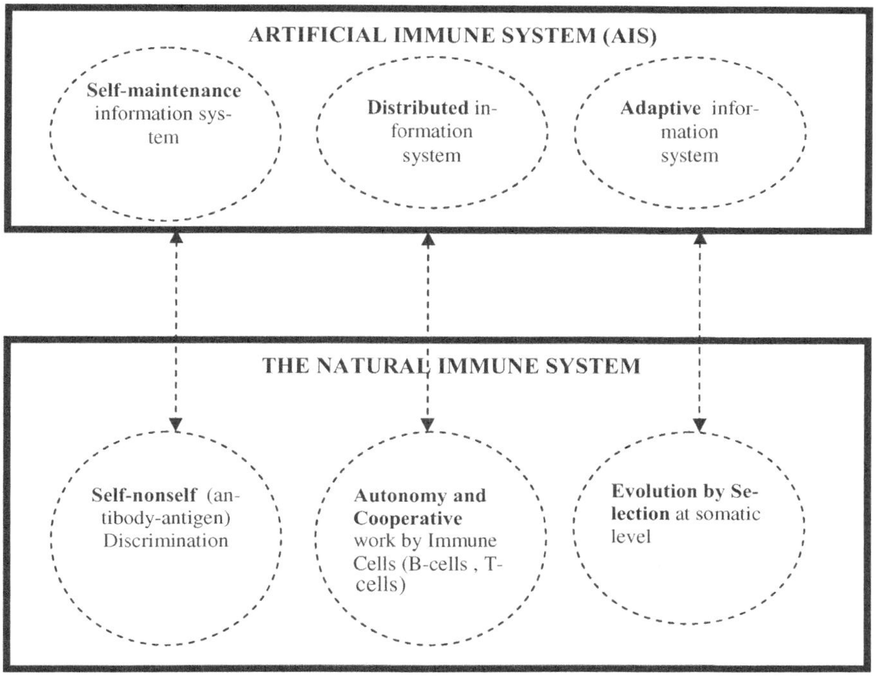

Fig. 1.2 A simplified block description of the relation between **AIS** as an information system and the Natural Immune System

The fundamental differences in information processing that make the distinction between an **AIS** and a usual information system are explained through a double suggestion both on the pattern recognition and regarding the processed data. It suggests that an ordinary Information System is limited to just a pattern recognition by the simply classification with classes, a mapping to a number of classes, whilst **AIS** performs in fact another kind of operation called dichotomy, that runs qualitatively different at a metalevel, namely from classes captured at the same level. The second message suggests that **AIS** incorporate the self-system and its relation to the outer world, but not at all prepare or embed a part of the solution into the model. Regarding the processed data, an **AIS** deals with challenges affecting the system itself, not with data that can be defined without referring to the system.

The basic elements of the framework that implements both the **AIS** structure and methodology involve: a representation (encoding) of **AIS** components; the evaluation (affinity) functions measuring the interaction between **AIS** components; typical algorithms managing **AIS** dynamical behaviour. The encoding must not only comprise just suitable elements corresponding to the antibodies or to the antigens, but also must reflect the strength of binding between an antigen and an antibody, also including other refined details such as the fact that an antigen can

STEP 1 Application Specification

1.1 Global Requirements Specification of the problem (task) to be solved

1.2. Detailed Specification of the Basic (Structural) Application Elements to be handled by the **AIS** (e.g., variables, constants, agents, functions, application specific parameters)

↓

STEP 2 Selection of the AIS Model(s) and Algorithm(s) Fitting the application

(Sometimes even more than one **AIS** may be developed for simultaneously acting to solve an application)

↓

STEP 3 AIS(s) model's Algorithmic implementation and run

3.1 Defining the Immune Components (the sets and types of antigens and antibodies)

3.2 Encoding (Representing) the Immune Components and selecting the corresponding type of Affinity Evaluation

3.3 AIS Algorithm Runs its appropriate dynamic behaviour (The concentration of antibodies must reflect a suitable match of the antigens with a set of antibodies associated with a set of appropriate behaviours)

3.4 AIS Metadynamics simultaneously related to Step 3.3 (insertion and elimination of antibodies from the network, possibly pre venting the alteration of **AIS** strength of action/creativity, by introdu cing a mechanism to generate candidate antibodies to enter the **AIS**)

↓

STEP 4 Real world implementation of the AIS solution

4.1 Decoding the ***AIS*** solution

4.2 Downloading the decoded (interpreted) solution into the application for real world

Fig. 1.3 The main steps of **AIS** Engineering Design

be a foreign antigen or a portion of an antibody, this means a self-antigen. The main steps of **AIS** Engineering Design look as in Fig. 1.3 for most of the application developments (Negoita 2005a), (Negoita 2007).

The artificial immunity-based systems are hybrid information systems almost by their nature as from the three main informational features so they offer an optimal technical frame of hybridisation with all other **C***omputational* **I***ntelligence* paradigms both at the level of model and most important at an algorithmic level (Negoita 2005), (Negoita and Reusch 2005).

1.2 Features and Classifications of Hybrid Intelligent Systems

HIS development application is justified in that no **CI** or **AI** method can be applied universally to every type of problem, each technique having its own advantages and disadvantages. It was the computational and practical issues of real world-applications, highlighting the strengths and weaknesses of **CI** and **AI** methods that led to evolution and development of **HIS**. The evolution of **HIS** s a consequence of modelling human information processing, but also it was the interdisciplinary vision of solving a large range of real-world problems that resulted in application engineering hybridising the **CI** and **AI** methods. **HIS** software engineering acts as a complex and complete social development agent, a so-called "business engineer" augmenting the **CI** information processing by the addition of typical knowledge elements to **AI**. So non-living systems are modelled on the biology of human intelligence by the integration of both **CI** and **AI** methods.

Most **HIS** display some *key features*, which make them particularly useful for problem solving. Certainly no **CI/AI** technique exhibits all these features. The key (application required) features making **HIS** development suitable for use with applications are as follows: *learning*, *adaptation*, *flexibility*, *explanation* and *discovery* (Goonatilake and Treleaven 1996).

An **HIS** must be able of performing a simple mining through a huge amount of previously acquired records of input data, to arrive at a model of the application. This **HIS** feature is called the *learning* ability and means the capability of learning the tasks/decision that the hybrid system has to perform, directly from the collected data.

An **HIS** must be able to express the intelligence of having decision-making procedures that execute in way that can be understood by humans. Decision-making procedures must be transparent, allowing the reasoning process to be understood and modified to improve the hybrid system. This **HIS** feature is known under the name of *explanation*.

Learning is a process that must continuously govern the activity of any **HIS** in any conditions (learning just an initial amount of knowledge is not at all enough for making a hybrid system able of performing a particularly task). **HIS** must be able to monitor the system tasks constantly (including knowledge revision according to any change in the operating environment). This feature is called *adaptation*.

The hybrid systems must be able to perform decision making even when there is imprecise, incomplete or completely new input data. This basic **HIS** feature is called *flexibility*.

Discovery is nothing else than **HIS** data mining ability – the capability of mining through a huge amount of collected input data and not only finding relationships that were previously unknown, but checking to see whether the discoveries were not just statistical flukes.

Accurate quantitative methods are not available for assessing the potential abilities of the main **CI/AI** techniques in providing the main desired **HIS** properties. A qualitative balance using the remarks accumulated and proved as a result of application development might perform the analysis, see Table 1.2 as from (Negoita , Neagu and Palade 2005).

Specific comments are to be made to a desired ideal implementation of the Adaptive Hardware Systems, namely for conferring them the typical **HIS** features of real-time systems. Real-time systems are systems that function under very sharp (application required) time and space constraints. For example, some of the usual time constraints include response/transition times and reasoning under time constraints. There are still a lot of practical issues that drastically limit the availability of the main **CI/AI** methods to provide **HIS** with the required real-time qualities (including high reliability). This aspect is more clear in case of a very well known class of adaptive **HIS**, the **EHW** namely.

Table 1.2 Comparison of main **CI/AI** techniques in providing the five main desired **HIS** properties

CI/AI Technique	*Learning*	*Explanation*	*Adaptation*	*Discovery*	*Flexibility*
KBS	inadequate	excellent	inadequate	inadequate	inadequate
FS	inadequate	moderate	inadequate	inadequate	excellent
NN	excellent	inadequate	excellent	adequate	excellent
GA	Excellent	moderate	good	excellent	good

Some of the issues and design limits of the **EHW** are discussed in (Negoita and Sekanina 2007) It is clear that greater technological support is needed. An active role with this aim is actually played by the embryonic/ morphogenesis hardware and the reconfigurable computing architectures, but also by the on-chip learning and adaptation with analogue circuits It is a realistic claim that the main **CI/AI** methods presently available are still not good enough to meet some applications requirements for special time, space and even sharp logical constraints. Severe consequences might result if logical as well as timing correctness properties of a real time system are not satisfied.

Table 1.2 has a double utility to the practitioners .It suggests two practical strategies in building **HIS** and also a classification according to these design strategies (Goonatilake and Treleaven 1996). This classification divides **HIS** into two classes: *function–replacing* **HIS** and *intercommunicating* **HIS**.

Function–replacing **HIS** are systems that are built by combining a **CI/AI** method that is inadequate with respect to one key **HIS** feature, with a method that is good to excellent with respect to that key feature. A lot of function-replacing **HIS** is already traditional. See for example: **GA-NN HIS** that are used for **NN** optimisation (**NN** weights and/or structure adjustments), (Arotaritei and Negoita 2002).

Intercommunicating **HIS** are systems relying on *technique-to-task allocation*: a complex application problem is sub-divided in specialised component problems (tasks), and a suitable **CI/AI** technique is allocated to each of these tasks. So different component problems, each of which may require different type of information processing, are solved by different specific **CI/AI** techniques and the final (global) results of the **HIS** are performed by communicating the specific results to each **CI/AI** techniques among themselves. An example may be the problem of launching a new manufactured product: a *forecasting task* is solved by using a **NN** technique, a *multi-objective optimisation* task is solved by using **GA** techniques and may be other reasoning tasks would use **FS/KBES** techniques.

An engineering classification point of view rather than a "systemic" one might be kept even if we are looking to theoretical (structural) aspects of information processing in **HIS**. In this case **HIS** are grouped into four classes: *fusion* systems, *transformation* systems, *combination* systems and *associative* systems (Khosla and Dillon 1997).

Fusion **HIS** is featured by a new **CI/AI** technique, *X*, whose procedure of information processing and/or its representation features are melted into the representation structure of a host **CI/AI** technique, *Y*. The practical aspects of melting a new **CI/AI** technique in a fusion **HIS** means that the host technique *Y* is diminishing its own weakness and exploits its strength more effective for solving the real-world application under these new circumstances. Let us mention some applications of **NN** based hybrid systems in which **FS** information processing features are fused: fuzzification of input data; optimisation of fuzzy systems (Fagarasan and Negoita 1995). Awell known **GA** based hybrid system in which **FS** information processing features are fused is the Lee/Takagi **GA** based system of intelligent integrated **FS** design, where both the membership functions and the number of useful fuzzy rules, as well as rule-consequent parameters are designed automatically and optimised using **GA** (Lee and Takagi 1993).

Transformation **HIS** are featured by the transformation of one form of information representation into another one. These hybrid systems are used for applications where the required knowledge for task achievement is not available and where one **CI/AI** method depends upon a different **CI/AI** method for its reasoning and processing purposes. An example may be a neuro-fuzzy system that makes use of learning fuzzy *IF-THEN* rules, learning fuzzy clusters and learning membership functions. Here **NN** learning features are integrated as a method operating on input-output data pairs or simply on input data.

A special case of transformation **HIS** occurs in special application circumstances where transformation is required to be in form of a complex optimisation. Firstly, an un-optimised technique *X* is melted into representation *Y*. Secondly, an extraction of the optimised prior representation *X* is performed from optimised representation *Y*. This is an intermediated class of **HIS** that is usually called a *fusion and transformation* **HIS.** Such systems are **GA-NN** or **GA- FS-NN** systems performing a complex **GA** optimisation of **NN** or **fuzzy NN**, namely envisaging the **NN** connection weights, **NN** structure, **NN** topology and **NN** input data. See (Arotaritei and Negoita 2002), (Palade, Bumbaru and Negoita 1998).

Another class of hybrid system is called *intelligent combination* **HIS,** *in* which two or more **CI/AI** technologies are combined. The distinct role played by each of the technologies in combination results in a more effective problem solving strategy. This combination **HIS** approach has an explicit hybridisation essence that relies on a refined matching of the **CI/AI** methods to the particular components of a modular structured model. The most common combination hybrid systems are **FS-NN-GA** hybrid systems for developing intelligent **GA** learning based adaptive control applications with **FS-NN** controllers. Combination systems perform a large variety of **KDD** specialised forecast/prediction tasks and also combination systems for complex tasks in robotics (**KDD** stands for **K***nowledge* **D***iscovery* in **D***atabases*).

No **HIS** developed can be classified as being a general-purpose technology capable of handling any application. All four classes of **HIS** have their limitations: the fusion and transformation systems - sometimes are not able to capture all aspects of human cognition related to solving the application; the combination **HIS** - seems to be the most complete system, but expresses a lack of minimal knowledge transfer among modules, despite their system flexibility; the range of tasks covered by fusion systems is restricted by loosing of the declarative aspects of solving the application (this is due to conversion of explicit knowledge into implicit knowledge).

The drawbacks of a fusion, a transformation, a fusion-transformation and a combination system favours the implementation of a more powerful system in solving applications, called *associative* **HIS**. Associative **HIS** may incorporate fusion, transformation and combination architectural/concept elements as well as integrate standalone **CI/AI** techniques. The results of these strategies confer a much-improved **HIS** quality features with respect to the quality of task achieved and to the range of tasks covered (Khosla and Dillon 1997).

Hybridisation meets application requirements and is strongly related to real-world problems of the Adaptive Systems. Hybridisation might be considered as an hierarchical process.

The first hybridisation level caters to applications of a low to moderate complexity and consists of just combining some **CI/AI** techniques to form an **HIS**.

The second hybridisation level is typical to applications of high complexity where some **HIS** is combined, together or not, with some stand alone **CI/AI** techniques. If we consider just the "silicon environment", these two levels of hybridisation are nothing more than a problem of systems adaptivity, namely an evolution of design strategies. However, it is now possible to look past simply the "silicon environment". Despite still being very much in the pioneering stage, quite promising steps have been made in a possible transition from the silicon environment hosted by microchips to the so-called "carbon environment " hosted by **DNA** molecules. Organic molecules containing **DNA** molecules intrinsically provide huge results of information processing capability. So the way is paved for a third hybridisation level performing hybridisation of some stand alone **CI/AI** techniques, even of some complete **HIS**s, along with the biological advances being made.

The third level of hybridisation means a bio-molecular implementation of soft computing, so that the uncertain and inexact nature of chemical reactions inspiring **DNA** computation will lead to implementation of a new generation of **HIS**. These are the so-called **R***obust* (**S***oft* **C***omputing*) **H***ybrid Intelligent* **S***ystems* – **RHIS** (Negoita , Neagu and Palade 2005).

RHIS are systems of biological intelligence radically different from any kind of previous intelligent system. The difference is expressed in three main features:

- *robustness* - conferred by the "carbon" technological environment hosting the **RHIS**
- *miniaturization* of the technological components at a molecular level the highest *(biological) intelligence* level possible to be implemented in non living systems
- dealing with world knowledge in a manner of *high similarity to human beings*, mainly as the result of embedding **FL**-based methods of **C***omputing* with **W***ords* and **P***erceptions* (**CWP**) featured by the understanding that perceptions are described in a natural language.

1.3 Emergent Intelligent Technologies and the Adaptive Hardware Systems: AIS – A Technology for the Adaptive Systems

There are some typical features that the Adaptive Systems display at the software level, which are featuring their behaviour in most of the applications. Their system behaviour is mainly featured by the capability to maintain or improve the performance in the context of internal or external changes. These changes may be, for example: uncertainties and variations during fabrication, faults and degradations, modifications in the operational environment, incidental or intentional interference, different users and preferences, modifications of standards and requirements, some compromises between performances and resources. At the hardware levels, Adaptive Systems certainly increase the system capabilities beyond what is possible with software-only solutions. There are a large number of hardware specific adaptation features employing both analogue and digital adjustments in the most elementary system components. Different algorithms, techniques and their implementations in hardware cover a large variety of applications, each of them providing a specific adaptation: adaptive communications – adapting to changing environment and interferences; reconfigurable systems on a chip and portable wireless devices – adapting to power limitations; survivable spacecraft – adapting to extreme environments and mission unknowns.

The use of biologically inspired **CI** techniques play a crucial role in developing robust and effective applications where complex adaptive systems set their face successfully against the large diversity of unpredictable and dangerous events that exploit the weak points or systems holes. Two emerging and promising biologically inspired techniques, **A***rtificial* **I***mmune* **S***ystems* (**AIS**) and **DNA** computing, seem to be the impulse of the moment in developing the strategy of systems survival in the defence of actual information systems against malicious faults (Ishida 2004), (de Castro and Timmis 2002). The collective effort of a large spectrum of high technology practitioners, mainly the computer scientists, engineers acting in different technical fields, biologists and natural environment specialists, led to the interdisciplinary development approach of **AIS** and **DNA** reliant hybridisation algorithms, techniques and applications.

One aspect of interdisciplinary interaction of biology, electronics engineering and computer science led towards research aimed to develop techniques involving information processing capabilities of biological system to supplement, and, perhaps, finally to replace the current silicon-based computers by so called **Q***uantum* **C***omputing* (**QC**) and **DNA** *Computing*.

DNA computing called also **M***olecular* **C***omputing* (**MC**) utilizes a framework of rules inspired by interaction of biomolecules and the protocols that govern microbiology. But developments are still far from being reliable and easy to implement in silicon-based computers. A software platform - *Edna* – has been developed to improve the encoding, reliability and efficiency problems of molecular computing (Rose et al 1999). An important and useful component of this software platform is the *virtual test tube* (a test tube simulator) giving practitioners a practical tool for a realistic test of any **DNA** computing protocol, before its implementation in a real Lab test tube. A general view on **DNA** computing hybridisation with other Intelligent Technologies may be found in (Negoita , Neagu and Palade 2005).

Artificial Immune Systems (AIS) are still considered with an attitude of reserve by most practitioners in **C***omputational* **I***ntelligence (***CI***)*, much more some of them even considering this emergent computing paradigm in an infancy stage. But in fact AIS are of real interest for the Adaptive Systems, this interest starting from the real world of applications that is asking for a radical change of the information systems framework. Namely, the component-based framework must be replaced with an agent-based one, where the system complexity requires that any agent to be clearly featured by its autonomy. The **AIS** methods build adaptive large-scale multi-agent systems that are open to the environment, systems that are not at all fixed just after the design phase, but are real-time adaptive to unpredictable situations and malicious defects. The **AIS** perform the defense of a complex system against malicious defects achieving its survival strategy by extension of the concept of organization of multicellular organisms to the information systems. The main behavioral features of **AIS** - as self-maintenance, distributed and adaptive computational systems - are well defined and suitably described in relation to the natural Immune System as an information system. A comparison of **AIS** methodology with other *Intelligent Technologies* is another point of the topic in this book. An overview of some actual **AIS** applications is made using a practical engineering design strategy that views **AIS** as the effective software with agent-based architecture.

Some basic considerations related to the natural immune system are to be made, but limited to just an introduction of the elements of immunology knowledge that are in connection with the defence mechanism of immune systems.

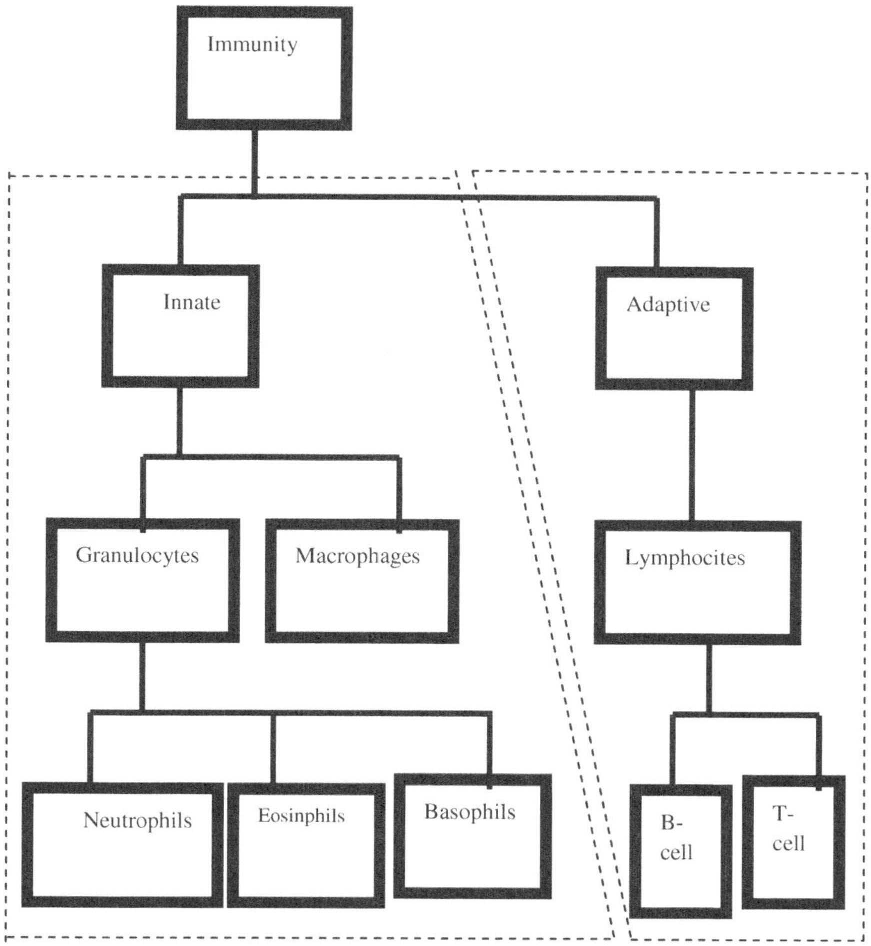

Fig. 1.4 A simplified block diagram of how the defence mechanism of the natural immune system is structured (after de Castro & Timmis 2002)

The natural immune system is a system of high complexity. Its physiology is featured by a bunch of spectacular and useful functions, among them being a highly effective defence mechanism for a given host against pathogenic organisms and infections. This defence strategy acts by performing two tasks: firstly, the recognition is achieved of all cells within the host body, namely whether they are self (belonging to the body) or nonself (not belonging to the body); secondly, the distinction between body own's cells and the foreign invader cells is followed by a classification of the nonself cells together with the induction of some appropriate defensive mechanisms for each of these dangerous foreign antigens that can be bacteria, viruses and so on. Details from different works in immunology science (Jeme 1973), (Jemme 1985),(Percus et al 1993) converge to a unique simplified block diagram of how the defence mechanism of the natural immune system is structured, see Fig. 1.4 after (de Castro and Timmis 2002).

A lot of interesting aspects regarding basic immune recognition and activation mechanisms, more deep details in physiology of the immune system, innate immune system and adaptive immune system and other functional fundamentals as pattern recognition, the clonal selection principle, self/nonself discrimination or immune network theory are to be mentioned when an overview is made on the basics of immunology, but this is beyond the aim of the book.

The book is focused on defense activity of the natural immune system that is achieved by the white blood cells, *leukocytes*, under a strategy of defense structured in a form of two distinctly implemented tasks of defense:

- the ***Innate Immune System*** (**IIM**) and
- the ***Adaptive Immune System*** (**AIS**).

The Innate Immune System (**IIM**) is implemented by two kinds of leukocytes, the *granulocytes* and *macrophages*. **IIM** combating responsibility consists of the fight against a wide range of bacteria without requiring previous exposure to them. Any is its body exposure to an antigen, the **IIS** response remains constant along the life time of an individual. A special combating strength features both the macrophages and the neutrophils: they are able of ingesting and digesting several microorganisms and/or antigenic particles; accordingly they are called together as phagocytes. But the macrophages are more powerful by having also the strength to present antigens to other cells. The granulocytes are cells with multiglobule nuclei containing cytoplasmatic granules filled with chemical elements (enzymes).

The following three kinds of the granulocytes are known, namely: the *neutrophils*, that are the most abundant **IIS** cells; the *eosinophils*, with a main task in the fight against infection by parasites; the *basophiles* with their functional task still not well elucidated.

The ***Adaptive Immune System*** (**AIS**) is implemented by one kind of leukocytes, the *lymphocytes* that are responsible both for the recognition and for suppression of a pathogenic agent. **AIS** combating responsibility is performed by producing antibodies just only in response to specific infections. This means that the presence of antibodies in an individual mirrors the history of all infections to which its body has already been exposed, either in case of a disease or a vaccination, and the practical action of **AIS** lymphocytes results in immunity against re-infection to the same infectious agent. This result proves the fact that **AIS** lymphocytes are capable of developing an immune memory. The lymphocytes are capable of recognizing the same antigenic stimulus at any time when it is presented again. The immune memory avoids the re-installation of the disease inside the body. Much more, **AIS** physiological mechanism improves the natural immune systems with each encounter of a given antigen.

There two main kinds of lymphocytes are as follows: *B-lymphocytes (B-cells)* and *T-lymphocytes (T-cells).* The **AIS** is acquired through the lifetime of an individual and is not at all inherited by an offspring. **AIS** acronym stands for Adaptive Immune System just in this paragraph, this notation stands for Artificial Immune System for all next sections of the book.

Two structural details of the natural immune system are crucial for both the terminology and comprehensibility of an **AIS**: the lueockocyte's receptor recognizing a particular molecular pattern is called an *antibody;* the molecular part recognized by an antibody is called an *antigen.*

Some elements of the **AIS** concept and framework are to be introduced starting from the conceptual philosophy of artificial immunity-based systems that revolutionizes the engineering of complex artificial intelligent systems.

Also a comparative discussion of the **AIS** against other biologically inspired *Intelligent Technologies* must run. This is required because the rest of branches in *Computational Intelligence* influenced the framework of **AIS** design, especially but not at all exclusively through **N***eural* **N***etworks* (**NN**) and **E***volutionary* **A***lgorithms* (**EA**), (Negoita, Neagu and Palade 2005).

The basic elements of the framework that implements both the **AIS** structure and methodology involve: a representation (encoding) of **AIS** components; the evaluation (affinity) functions measuring the interaction between **AIS** components; typical algorithms managing **AIS** dynamical behaviour. The encoding must not only comprise just suitable elements corresponding to the antibodies or to the antigens, but also must reflect the strength of binding between an antigen and an antibody, also including other refined details such as the fact that an antigen can be a foreign antigen or a portion of an antibody, this means a self-antigen. The main steps of **AIS** Engineering Design look as in Fig. 1.3 for most of the application developments (Negoita 2007, Negoita 2005a).

AIS versus other techniques of *Computational Intelligence* must be discussed starting from the idea that the algorithms managing **AIS** dynamical behaviour can be applied to problem solving in different settings. But an essential detail of **AIS** engineering states that the pool of immune algorithms is mainly divided in two categories: **p***opulation-***b***ased* (**PB**) and **n***etwork-***b***ased* (**NB**) ones. We just limit ourselves here by mentioning the fundamental feature that divides **AIS** into the above mentioned two classes: the encoded components of **PB-AIS** interact exclusively with the external environment – represented by antigens, this is the case of bone marrow or thymus models; the encoded components of **NW-AIS** are more co-operative by interacting both with each other and with the antigens representing the external environment, this is the case of immune network models.

This book will focus not so much on the similarities and differences among **AIS** and other biologically inspired **I***ntelligent* **T***echniques*, see (Forrest 1990) for a comparison between **AIS** and the **GA,** or (de Castro et al 2000), (Dasgupta 1997) in case of **AIS** versus **NN.** It will be useful to make a briefly overview of the possible **AIS** aggregation with the other *Computational Intelligence* approaches in form of **HIS** that influenced both **AIS** design and their applications (Negoita, Neagu and Palade 2005), (Negoita 2005a).

The artificial immunity-based systems are hybrid information systems almost by their nature as from the three main informational features as previously mentioned.

An example of **AIS-NN** hybridisation at the level of model may be found in (Hoffman 1986) where a new **NN** learning algorithm was built. The learning is performed by strength variation of the input stimuli instead of varying the **NN** weights that are constant. Also the **NN** memory capacity was increased due to the **AIS** behaviour as a system with a large number of attractors.

The great potential of **AIS-EA** hybridisation is illustrated by a large variety of technical improvements to most of the **EA** paradigms. Some bidirectional improvements featuring the **AIS-GA** hybridisation were reported: **AIS** niches, species and diversity were controlled by a **GA** of the hybrid immunity-based algorithm in (Forest et al 1993); **GA** constraints handling was made by a simulated **AIS** where the antigens fight against the antibodies, so as the resulting antibodies are the constituents of an evolved population of constraint conditioned individuals (Yoo et al 1999).

Even an **AIS**-based **GP** variant that uses an **AIS** dynamic fitness function was reported in (Nikolaev et al 1999).

FS – **AIS** hybridization was strongly used in a wide range of real-world applications. Distributed Autonomous Robotic Systems may have a **FS** like modeling of the stimulation level of an antibody (individual robot strategy of action), while an ***AIS*** relying on clonal selection is used for transmitting high quality strategies of action (antibodies) among the robots. No central control exists and the role of antigens is played by elements of external environments (Jun et al 1999).

Other advanced **AIS**-based **HIS** will be introduced in the next paragraphs, including **AIS** aggregation with the emergent *Intelligent Technologies* such as **E***volvable* **H***ardware* **(EHW)** or **DNA** *Computing*.

Multi-agent framework is the adequate environment for **AIS** Engineering. The fundamental design philosophy for any complex information system must definitely escape from the *component-based framework* to the *agent-based framework* where each agent has its own intelligence and autonomy in order to attend the required complexity (Ishida 2004). An agent is defined as any entity – human or software- capable of carrying out by itself an activity standalone.

Using the **CI** methods creates an agent-based framework where the own intelligence of each agent may even evolve to a behavioral one. A multi-agent approach using an arsenal of **CI** methods that is mainly relied on the **AIS** would perform the defense of a complex system against malicious defects and other unpredictable events, achieving its survival strategy by extension of the concept of organization of multicellular organisms to the information systems. The main reasons of using the agents for information processing and managing are as follows: the agents are proactive, they are cooperative, and they are able of learning and also of reasoning, as the case (Vizcaino et al 2003).

The agents are proactive, means they initiate by themselves the decision making for an action when they find it necessary to proceed in such a way. This task is crucial because the information may be generated by different sources and often from different places. The agent's ability of cooperation/information interchange means their availability of knowledge sharing among them or benefiting from the other agents' experience by asking for their advice. By learning agents we mean just agents that can learn from their own previous experience, comprising both mistakes and successes. Finally, each agent in the **CI** multi-agent framework may utilize any sophisticated **CI** technology of reasoning.

The infrastructure of a multi-agent system includes: agents, agent platforms, agent management and an agent communication language (Tianfeld 2003). The multi-agent approach to complex systems enables the separation of social behaviour (problem solving, decision making and reasoning) at social communication level from individual behaviour (routine and knowledge processing, and problem solving) at autonomy level. The two distinct communication levels, the social one and the autonomy level, along with the dynamic agent-task association, frame the multi-agent infrastructure.

The **AIS** main informational features - the *self-maintenance* property, the property of a *distributed system* and the property of *an adaptive system* justify a continuous elaboration work of different variants of **AIS** agent-based architecture (Ishida 2004), (Okamoto et al 2003), (Ishida 1996b), (Watanabe et al 2004), (Okamoto et al 2004).

The so-called "most naïve immune algorithm" was introduced in (Ishida 1996b) as a proceeding to implement an adaptive **AIS** running in three steps, see Fig. 1.5 after (Ishida and Adachi 1996):

step 1 - Generation of Diversity;
step 2 - Establishment of Self-Tolerance and
step 3 - Memory of Non-Self.

The algorithm views the system as the "self" and the external environment as the "non-self". Both the self and the non-self are unknown or cannot be modelled.

Step 1 generates the recognizers' diversity in its specificity.

A developmental phase drives structural changes during Step 2 on the recognizers aimed of becoming insensitive to known patterns (self).

Parameter changes are driven during Step3 on the recognizers aimed to be more sensitive to unknown patterns (the non-self). The recognizer is an **AIS** unit featured by only recognizing and communication capabilities.

From a practical, engineering perspective, the Natural Immune System is a robust system and similar mechanisms must be implemented in the **AIS** to build robust systems for different applications. The system-level adaptability in an **AIS** is attained by diversity and selection, an adaptation that is inspired from the Natural Immune System.

The corresponding Agent-based architecture handles high performance intelligent and autonomous units called *agents* that beside the recognizing and communication capability are able of adaptation and self-replication.

The main three steps of the agent – based **AIS** have similar functional meanings to the immune system – based AIS: Step1 – achieves diversity generation; Step 2 - performs the activation of the recognizing agents by an encounter with the antigens; During Step 3, the activated recognizing agent will reproduce its clone to enhance the ability of elimination of the antigen.

Mutation operators have a role to increase the affinity with the antigen perform this reproduction. **AIS** agent-based systems must be technically specified as follows: systems designed for solving self-nonself discrimination problems (**AIS** are self-referential systems because the self is related to the **AIS** itself) and the discrimination runs as a consequence of the interactions among the agents. If the system is open to the external environment, the interactions with exogenous nonself are involved to achieve the adaptability by selection from the exogenous nonself.

The **AIS** agent-based framework uses agents (cells) featured as follows: homogenous in structure and potential, butt still specialized in function; able of reproduction under a slight mutation (typically, a point mutation).

The agents are able of changing their algorithmic parameters (mutation rate, lifespan, reproduction rate) when triggered by special events.

An interesting fault tolerant **AIS**-based multi-agent architecture of a distributed system was proposed that performs self-repairing (Watanabe et al 2004). The performances of this architecture were proved by a distributed computer network system consisting of *N* host computers each of them being able of sending mobile agents to adjacent hosts. The abnormal units, either host computers or mobile agents are identified by an **AIS** strategy. Some units try to self-repair; by this means these units replace their data with data received from other units.

Some other **AIS** applications in context with the **HIS** framework are also connected to the topic in this book, namely that ones progressing toward some spectacular and effective combinations envisaging even the emergent **CI** paradigms such as **DNA** *Computing* and **EHW** (Negoita , Neagu and Palade 2005), (Negoita 2005a).

A **DNA-AIS** intelligent hybrid system was reported in (Deaton et al 1997), where **DNA** *Computing* was proved as an alternative to implement **AIS**. In this work, an **AIS** negative selection algorithm was implemented in a **DN***A* computing framework. Using **DNA** single strands under denaturation, renaturation and splicing operators, the censoring and monitoring parts of this selection algorithm was successfully implemented.

A recent intelligent hybridization of **AIS** is applied in case of one of the most revolutionary technology nowadays, namely in case of **E***volvable* **H***ard***w***are* *(***EHW***)*. A main reason for **EHW** – **AIS** hybridization was reliant on two **AIS** features, healing and learning, that were applied to design **EHW** fault – tolerant **FPGA** systems (Bradley et al 2000). An additional layer that imitates the action of antibody cells was incorporated to the previously elaborated embryonic architecture by the same team (Ortega et al 2000). Two variants of this new **EHW** architecture use an interactive network of antibody cells featured by 3 independent types of communication channels: the data channels of the embryonic array of cells, the data channels of antibody array of cells and the inter-layer communication channels ensuring that antibody cells can monitor the embryonic cells. The

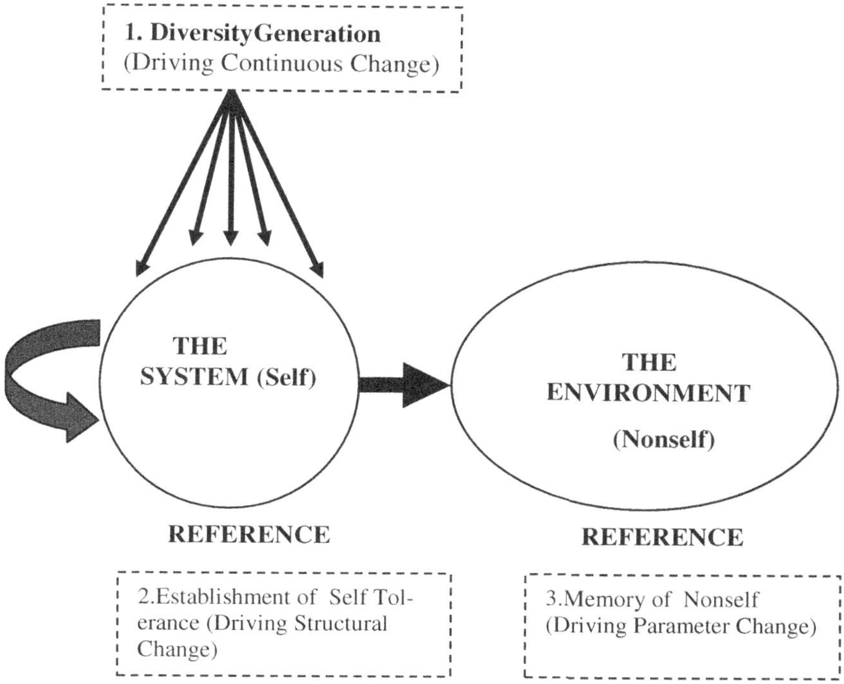

Fig. 1.5 AIS diagram, after (Ishida and Adachi 1996)

antibody array of cells performs monitoring and checking of the embryonic array of cells, so that the correct functionality of any particular **EHW** configuration can be achieved at any time.

Another **AIS** inspired variant of **EHW** hardware fault detection was reported in (Bradley and Tyrell 2001). They used an advanced **FPGA** hardware- Virtex XCV300 - to implement a hardware negative clonal selection **AIS** attached to a **F***inite* **S***tate* **M***achine* (**FSM**). This is very important because any hardware system can be represented by either a stand-alone or an interconnected array of **FSM**s.

The main **AIS** applications have been developed in areas such as: autonomous navigation/robotics, computer network security, job-shop scheduling, (fault) diagnosis, data analysis and optimisation (de Castro and Timmis 2002). Some of these applications are reliant on the idea of combining **AIS** with different other **CI** techniques (**FS**, **NN**, **EA**, **KBES**, **DNA** Computing) with the aim of creating **HIS** that are collecting the individual strength of each **CI** component (Negoita , Neagu and Palade 2005). **AIS** are a new **CI** approach that has not only applications involved in the **HIS** framework but also has its own standalone applications.

Refined details of behaviour arbitration for autonomous mobile robots were solved in (Ishiguro et al 1996) through new **AIS** based decentralized consensus-making system. The adaptation mechanism for an appropriate arbitration uses reinforcement signals that were used for evolving the proposed **AIS**: current situations detected by sensors work as multiple antigens, the prepared (desired) competence modules work as antibodies, the interaction between modules is represented by stimulation and suppression between antibodies.

But **AIS** applications have evolved so far that they are able to improve the lives of human beings even in the most unexpected aspects. **AIS** are also applied to improve the comfort in our daily real life at a micro level. The **AIS**-based smart home technology is aimed to improve the technical facilities, including the security of a home not only against the burglars but against other external dangers too. Such an **AIS** was modelled as a multi-agent system where the sensors and actuators are taken as agents that prove to behave as the human immune system with respect to self organization and flexible reaction, with gradation, on dangerous events outside (Dilger 1996).

But the **AIS** may act for improving our daily comfort even at a macro level. Potential applications of the **AIS** in some selected areas of physical infrastructure assessment and modelling at the national level, in particular for surface transportation (highways, railroads, air transportation facilities) are suggested in (Attoh-Okine 1996). For example, an **AIS** relied on negative selection is proposed for condition assessment by analysing the instances of a more general problem distinguishing itself (normal condition data not below a threshold value) from other (deteriorated data below the threshold). Also an **AIS** network model may implement the fault states diagnosis of infrastructure systems where each infrastructure subsystem is regarded as a distinct antibody, and the information about the state of the global infrastructure system is treated as an antigen. An **AIS**-based robotic system for infrastructure assessment may be applied for the maintenance activity in pavement infrastructure systems: the maintenance-actions are acting as antibodies, whilst the infrastructure conditions behave as antigens.

Most application areas in the major field of pattern recognition actually make use of **AIS**-based methods. Among them is the feature extraction in recognition of complex characters, such the Chinese characters (Shimooka et al 2003). An **AIS** model relies on the effect of diffusion of antibodies, namely the amount of diffused antibodies is calculated by adopting the (spatial) distribution of antibodies centroids as the virtual points where antibodies were concentrated and redistribution of antibodies is performed too. **AIS** based on the partial template method are effective for the personal identification with finger vein patterns (Shimooka et al 2004).

Security systems for computers and the Internet work environment are another productive application area of the **AIS**. Intrusion detection on the Internet – for internal masqueraders mainly, but for external ones too – is reported in (Okamoto et al 2003). The method was inspired by the diversity and specificity of an immune system, namely each immune cell has a unique receptor featured by a high degree of matching a specific antigen. So each of the agents that are used in this approach has its unique profile and computes a high score against the sequential (command) set typed by the specific user matching this profile. By evaluating all the scores (for all the profiles), one of the agents determines whether a particular user is an intruder (masquerader) or not.

Usually computer protection is referring to anti-virus protection and to intruders' detection, but (Oda and White 2003) applied **AIS** for immunity to unsolicited e-mail (spam or junk mail) where regular expressions, patterns that match a variety of strings, are used as antibodies. These regular expressions are grouped in a library of gene sequences and in their turn the regular expressions are combined to randomly produce other regular expressions to produce antibodies that match more general patterns.

The **AIS**s were proved to be superior to hybrid **GA** in function optimsation (Kesey and Timmis 2003). Here the **AIS** algorithm inspired by clonal selection and called **BCA** (**B**-**c**ell **a**lgorithm) got a high quality optimisation solution by performing significantly fewer evaluations than a **GA**. A unique mutation operator was used – contiguous somatic hyper mutation – that operates by subjecting the contiguous region of the operative element (vector) to mutation. The random length utilised by this mutation operator confer to the **BCA** individual the ability to explore a much wider region of the affinity (fitness) landscape than just the immediate neighbourhood of an individual.

A hybrid clonal selection **AIS** was used more successfully than the evolutionary algorithms for solving a combinatorial optimisation application, the Graph colouring problem (Cutello et al 2003). Here the use of a crossover operator was avoided by using a particular mutation operator combined with a local search strategy. In this way there was no embedding specific domain knowledge.

AISs that are based on clonal selection have proven to be effective both for combinatorial optimisation and for machine learning problems. **AIS** modelled as a noisy channel has applicability in adaptive noise neutralization (Cutello and Nicosia 2003a). Here the signal is the population of B-cells, the channel is the global **AIS**, the noise source is the antigen and the received signal E is the antibody.

Regarding the machine learning methods that are used in implementing classification algorithms for "hard" applications as the gene expression of cancerous tissues, **AIS** proved to be superior to **SVM** and **NN** by a better error rate of the algorithm, despite of a longer amount of computational time (Ando and Iba 2003).

Indeed, the **AIS** conceptual philosophy revolutionizes the engineering of adaptive hardware and software intelligent systems by extension of the concept of organization of multicellular organisms to the information systems in a crucial way: the **AIS**-based modern complex information systems are adaptive large-scale multi-agent ones, that are open to the environment, they are not at all fixed just after the design phase, but are real-time adaptive to unpredictable situations and malicious defects. **AIS** offer the unique framework performing the defense of a complex adaptive hardware or software system against malicious defects achieving its survival strategy.

AIS application area has become already traditional and accurate in fields such as: computer and data security, image and other pattern recognition (including signature verification), fault and anomaly detection and diagnosis, noise detection, autonomous navigation and control, scheduling, machine learning, machine monitoring. The **AIS** feature some very effective general-purpose engineering and science design methodologies with a large applicability for search and optimisation strategies and also for design of the adaptive multi-agent based systems.

AIS actually penetrate some totally new and unconventional application area of our social-economic life, such as the smart home technology. **AIS** confer a highly effective adaptative feature to the modern instructional and educational tools of **I***ntelligent* **T***utoring* **S***ystems* (**ITS**). These systems model instructional and teaching strategies, empowering educational programs with the ability to decide on "what" and "how" to teach students. A "stand alone" intelligent, HIS reliant, tutoring component is added to the usual learning environments, so the work that is done by lecturers and students is complemented (Negoita and Pritchard 2003), (Negoita and Pritchard 2004). Usually, learning systems require that the student change to fit the system, but the *ITS* radically differ by an added flexibility of performing an intelligent learning strategy, namely the learning system *"change to fit the students needs"*.

AIS proved to be an effective standalone **CI** paradigm for many engineering applications including computer science, but it strength is sensitively increased in aggregation with other **CI** paradigms in form of **HIS**. **AIS**-based **HIS** are applied in the area of hardware/software adaptive systems as distribute autonomous robot systems, classification and prediction systems of high performance, or risk analysis evaluations featuring highly complex applications as the highway infrastructure deterioration at the national level.

As was mentioned previously in this book chapter, **AIS** are used to distinguish self (normal) from nonself (abnormal or attack). The **AIS** self is a set of entities which are part of the monitored system. The **AIS** nonself is a set of foreign entities that may damage the monitored system. By definition, the expected **AIS** reaction must be just to nonself, the reaction to any self-part of the system is not desired. The same happens inside the biological immune system: the populations of antibodies are trained to distinguish self from nonself, with the main aim of preventing system reactions to self.

AIS-GA based **HIS** can use **GA** for improvement of training the antibodies to detect the nonself attacks. An attack may be viewed as having a malicious content consisting of certain characters or character strings that are also featuring a normal content. These characteristics may be used as a fingerprint to see if the content is a malicious or a normal one. The **AIS** doesn't use a signature for attacks, but just its own principles for attack detections: random generation, negative selection, partial matching and affinity maturation.

The antibodies are randomly generated at the start of the algorithm, but during this step some antibodies that would react to self can be produced. The negative selection plays the role to destroy these not desired antibodies. The process of producing just antibodies that only react to nonself is a combined action of the negative selection with a partial matching. The negative selection performs as follows: a representation of self is presented to the antibody; if this antibody reacts to self , it is destroyed. But the self representation may be not complete , so a generalization must be achieved for a suitable reaction of the antibody. The generalization task is performed through the partial matching. An intrinsic **AIS** weakness upsets the initial random population of antibodies by not performing enough well at detecting nonself, so affinity maturation intervenes for improving the performance of antibodies. The **AIS** weakness is counterbalanced by the **GA** (optimisation) strength: the **GA** is used to breed the population over several generations to optimise the detection ability of the population of antibodies. This kind of **HIS** was successfully applied in detection of Web server attacks (Danforth and Levitt 2003).

AIS play a key role in behaviour-based **AI**, which proved its robustness and flexibility against sharp dynamics of the changing external world. This kind of **HIS** intelligence is the result of both mutual interactions among competence modules (behaviour/action modules) and interaction between a robot and the external environment. An important issue of the above-mentioned interactions, is to skip a fixed priority basis of arbitrating the competence modules – as in case of using subsumption architectures (Brooks 1991). Another flexible interaction mechanism -the behaviour network system - is difficult to be applied if the application is difficult to find the cause-effect relationship among the agents (competence modules). These two mentioned issue in robotic are successfully solved by the **AIS**.

AIS–based methods of behaviour arbitration for the autonomous mobile robots take into account the concept of *self-sufficiency* in order to implement a decentralized consensus-making bio-inspired system (Ishiguro et al 1996). Here a robot called "immunoid" has the task to collect garbage and put it into a garbage can without running out of his energy - his battery level The fact that the robot consumes some energy as it moves around the external environment, is looked like the metabolism in our biological system. The detected current internal/external situation works as an antigen, whilst the robot's prepared simple behaviour works as an antibody. The antigen informs about: the garbage direction (front, right, left, back); the obstacle direction (far, middle, near); energy battery level (high , low); home base (direction and distance). The appropriate selection of antibodies is called the concentration of antibodies. It denotes the affinities between two different antibodies, and between each antibody and the detected antigen. The adaptation mechanism aimed to perform an appropriate arbitration is relied by using reinforcement signals, this means the immunoid is a **HIS** of **AIS**-**NN** type.

Chapter 2
Advanced Hardware Implementation of the Computational Intelligence and Intelligent Technologies

A briefly overview on biological sources of inspiration for softening (soft) hardware is thought as useful for an easy understanding of **EC** in **EHW** implementation (Sanchez et al 1997). Natural Life on our planet just from its early beginning was structural organized in three distinct levels: *phylogeny* level – governing the temporal evolution of genetic programs within individuals and species; *ontogeny* level - the guidance of development process in single multicellular organisms (successive division of mother cell, followed by daughter cells specialisation or differentiation); *epigenesis* level – the learning process featuring the whole lifetime of an individual organism (nervous system, immune system, endocrine system are defined by the genome which is subject to modification through interactions of the individual with the environment). In analogy to Natural Life, **EC** is the artificial intelligence homologous part of natural *phylogeny*, **self-reproducing automata** are the homologous part of natural *ontogeny* and *neural networks* (**NN**) have an analogue behavior to epigenetic natural process in organisms.

The bio-inspired systems can be built and classified inspired by the powerful natural examples as following the above-mentioned levels of organization in Natural Life. The same considerations are applicable to **EHW** that may be viewed by its definitions as a sub- domain of artificial evolution.

The *phylogenetic hardware* is of different types:

- *evolutionary designed circuitry* (all operations are performed in software; the resulted solution is possible to be downloaded into a real circuit)
- *real circuitry evolution* (a real circuit as a hardware environment evolved by operation *most* performed in software)
- *online evolved hardware* (a not open-ended evolution performed in hardware – an evolution with a pre-established goal and no dynamics)
- **EHW** (a population of hardware individuals evolve with in an open-ended evolution)

The *ontogenetic hardware* is illustrated by self-reproducing hardware. Different self-reproducing hardware methods were proposed: most known in the last years is *embryonics* (Marchal et al 1994) a replica of asexual multicellular living beings,

M.G. Negoita, S. Hintea: Bio-Inspired Tech. for the Hardware of Adaptive Sys., SCI 179, pp. 27–81.
springerlink.com

in fact a multicellular automaton based on multicellular organisation, cellular differentiation and cellular division, with self-repair capabilities too.

The *epigenesis* hardware tries to implement the architecture of natural brain by evolving **NN**s. A genetic encoding of **NN** structure is used for neural structures based on **C**ellular **A**utomata (**CA**) that grow (evolve) in a special hardware environment (Gers and de Garris 1997).

After the shocking emergence of **EC**, it followed a two-decade period of impressive theoretical studies and developments in the computer laboratories of the academic institutions. But now is going on its most benefit stage of development by loosing the theoretical results in the real-world, with multiplication and adaptation to a wide variety of industrial and commercial environments.

EC application success is huge indeed, because they seem to provide better solutions to a wide variety of complex optimisation, design, routing and scheduling problems.

The business world employs **EC** techniques in financial engineering both to improve the business fitness and ensure its survival in competitive markets.

On the other hand a lot of industrial applications are the result of **EC** techniques applied by a few companies throughout their operations (Mellis 1996), (Robinson 1996), (Kelly 1996), for example:

- aerospace applications – functional layout of switching matrix, attitude determination of a spacecraft
- design of antenna and electronic circuits or devices (including transducers)
- generation of bespoke solution to individual customer requirements in electronics industry.

Nevertheless, the most benefit for the society and indeed most revolutionizing application of **EC** is its hardware implementation leading to the **EHW**. These new **EC** based methodologies led to a new type of machines that is evolved to attain a desired behaviour, which means they have a *behavioural computational intelligence*.

What was far years ago a dream of technology became nowadays a reality: adaptation transfer from software to hardware is possible by the end. The electronics engineering was radically changed as a profession practically; there are no more technological limits in hardware circuits.

2.1 Evolvable Hardware: An Overview

A *definition of* **EHW** may be as follows: a sub-domain of artificial evolution represented by *a design methodology* (consortium of methods) involving the application of **EA** to the synthesis of digital and analogue electronic circuits and systems. A more agreed definition among the practitioners might be: **EHW** is *programmable hardware* that can be evolved (Toresen 1997). But some members of the scientific community acting in the area consider the term *evolutionary circuit design* more descriptive for **EHW** features. Much more, another term used nowadays for the same work is *evolware* concerning to this evolvable ware with hardware implementation, leading to a future perspective of using the term *bioware* concerning to a possible evolving ware with biologic environments implementation. Even other environments are seen as possible evolvable media: *wetware* – real chemical compounds are to be used as building blocks or *nanotechnology* – relied on molecular scale engineering.

This new design methodology for the electronic circuits and adaptive systems is not a fashion. It is suitable to special uncertain, imprecise or incomplete defined real-world problems, claiming a continuous adaptation and evolution too. An increased efficiency of the methodology may be obtained by its application in the soft-computing framework that means in aggregation with other intelligent technologies such as **FS** and **NN**, evolutionary algorithms (**EA**), **AIS**.

The reason of **EHW**'s using in the above mentioned type of application is its main advantage over the traditional engineering techniques for the electronic circuit design, namely the fact that the designer's job is very much simplified following an algorithm with a step sequence as below:

STEP 1 (*problem specification)* - requirements specification of the circuit to be designed

- specification of basic (structural) elements of the circuit

STEP 2 (genome *codification)* - an adequate (genotypic) encoding of basic elements to properly achieve the circuit description

STEP 3 (*fitness calculation)* -specification of testing scheme used to calculate the genome fitness

STEP 4 *evolution (automatically generation of the required circuit)* -generation of the desired circuit

The designer himself is involved by acting directly during the first three steps, while the fourth step is *automatically generation* of the circuit. The flow modality of both step *3* and step *4* leads also to same categorizing classes criteria for **EHW**.

Despite of the above-mentioned advantage of **EHW** methodology, its disadvantage is not to be neglected: the obtained results are (sometimes) suboptimal over those ones of the classical methods that remain preferable in case of well-defined problems.

EHW definition and terminology in context of the adaptive hardware systems was introduced in (Stoica and Radu 2007), namely having a starting point in definition of an **A***daptive* **S***ystem* (**AS**) and in definition of the **A***daptive* **H***ardware* (**AH**).

AS is defined as a system featured by the capacity to modify itself in order to maintain or improve its performances towards an objective, and/or in response to the changes or perturbation of the environment. A partial or a full system change is usually performed inside the **AS** by a controller that operates the change and an objective function that guides this change.

AH is viewed as a physical sub-system featured by the capacity to maintain or improve its performance towards an internal objective and/ or in response to a changing environment, either external or internal, as the case. An **AH** must consist compulsory of the following parts: the hardware that can change; the controller that changes the hardware; the objective function -the component part that guides the change. The objective function can be implemented by a built-in method, can be passed on by other distinct hardware or can be passed on by the user.

An **EHW** definition starting from the **AS** and **AH** states it as a physical sub-system featured by the capacity to modify itself with the aim to improve its performance towards an internal objective and/or in response to a changing operating environment. This means nothing else than a special case of **AH**, namely able of being driven by internal objectives mainly and featured by a continuous improvement of its performances over the time.

2.1.1 EHW Classification, Practical Engineering Remarks

Currently the common features of the adaptive hardware at different levels - adaptive materials or adaptive analogue circuits, adaptive digital circuits, reconfigurable computers, reconfigurable networks – are not well documented, and the same situation is in case of criteria of its classification. Both theoretical and application related criteria led to different kind of **AH** classification.

The **EHW** methods may be classified from the *genome-encoding* point of view in two classes (Tomassini and Sipper 1997):

- methods using *high-level languages* to encode the circuits. It means the final (genome) solution is transformed by a decoding equivalent operation to obtain the actual circuit (phenotype solution)
- methods using *low-level languages* to encode the circuits . It means the bit string representing the genome (codification of the circuit's basic logic gates and their interconnections), is possibly directly placed in the actual circuit, obtained without to transform the genome (without decoding).

There are two main **EHW** classes from the *fitness calculation* point of view:

- *offline* **EHW** – the fitness of a genome represented in a high-level language is calculated by simulation, started with a transformation (decoding equivalent operation) of the encoded solution; only the final solution as a result of simulated evolution is implemented in hardware
- ***online*** **EHW –** the fitness of a genome represented in a low-level language is calculated by achieving a direct reconfiguration of the circuit, as a result of evolution implemented in real hardware.

But a point of view relied on how an application runs, led to four **EHW** classes. The first class of **EHW** is called *evolutionary circuit design* with the following dynamics:

- carried out **EC** operations in software ("offline")
- the final resulted solution is downloaded into a real circuit

The second class of **EHW** may be called *online* evolution because of the following reasons:

- (most) **EC** operations are carried out "offline" in software, by an external computer hosting the population of individuals (genomes)

too

- a real circuit is still used during the evolutionary process
- the real circuit system is used with a sequential evaluation of every individual taking on it, at each evolutionary generation.

There are two practical solutions for implementing this type of **EHW** as follows (Negoita and Stoica 2004):

- the variant with *one real circuit* – a sequentially downloading of each individual in the offline population is performed; no real time behaviour can be achieved onto the real circuit before the evaluation process was ended
- the variant with *two real circuits* having two distinct tasks – one circuit executing the real time configuration of the best previously evaluated individual, the second circuit being devoted to fitness function calculation; this variant hasn't the "online" attribute really; the adequate call should be *quasi-online* **EHW** better.

The *third class* is denoted *online* **EHW**:

- all typical **EC** operations (selection, crossover, mutation, and fitness evaluation) are carried out online in hardware
- a predefined goal (global function) features this **EHW** leading to a not open-ended evolution.

The *fourth class* is denoted *truly* **EHW** and its behaviour is featured by a population of hardware individuals (circuits) that evolves having an open-ended evolution. But in opposition with the natural open-ended evolution that is undirected, here is a flexible directed open-ended one. The fitness function externally imposes.

This means that the user imposes the fitness criterion externally, as a consequence of the task to be solved. An **AH** classification starting from the type of digital functions changing via digital switches/ multiplexing was made in (Stoica and Radu 2007) and gives the five classes.

Analogue hardware that crosses quasi-instantly and directly from the initial to the final configuration by switches /multiplexing is called *reconfigurable analogue hardware*.

The analogue hardware that changes gradually without switches is called *morphable analogue hardware*.

Another class of hardware is featured by the fact that changes influence the parameters of a function. This hardware is known as *adjustable/tunable/parametric hardware*.

But a more generic **AS** classification – including **AEH** too - takes into account how the adaptive behaviour runs as a result of the Objective Function. This action was done by **EUNITE** (*the* **E***uropean* **N***etwork on* **I***ntelligent* **T***echnologies for* **S***mart* **A***daptive* **S***ystems)* (Leiviska 2004) and looks like as follows:

- o *External adaptation* **AS** – adaptive behaviour is performed in the presence of stimuli originating in the external environment
- o *Internal adaptation* **AS** – adaptive behaviour is performed in the presence of disturbances located inside the system itself
- o *Darwinian adaptation* **AS** – adaptive behaviour is performed by a response that is directed toward modifying the object
- o *Singerian adaptation* **AS** – adaptive behaviour is performed by a response that is directed toward modifying the external environment

The **AEH** practitioners must have a very accurate image of what a **S***mart* **A***daptive* **S***ystem* (**SAS**) means. At the first glance, " smart" tends to be understood in the frame of **CI** and **HIS**. This means **IT** included (**FS**, **NN**, **EC**, **AIS**, **DNA**, mainly embedded). The **EUNITE** community of specialists starts **SS** definition from the properties of the system. "A smart system has a model of its own behavior and not just a model of its environment. This means it does not only react to its environment, but it is also capable of predicting the repercussions of its own actions. It can predict how the environment will change and how that change will influence itself. It can use the results from its predictions to draw consequences for its own actions."

As a consequence, **SAS** is identified by **EUNITE** using a three level definition, namely a **SS** that can adapt to:

- to a changing environment (adaptation level 1)
- a similar setting without explicitly being "ported" to it (adaptation level 2)
- adapt to a new/unknown application (adaptation level 3)

Changing environment concerns with systems that adapt their operation in changing environment by using their intelligence to recognize the changes and to react accordingly. The second level considers the change in the whole environment and the system's ability to respond to it. Adaptation to a new/unknown application is the most demanding one and it requires tools to learn the system's behavior from very modest initial knowledge.

It is highly desirable that any real-world application of adaptive or evolvable hardware to achieve a performance that is featured both by technological benefits and by improvement in economics.

The technological benefits mean either /and/or better solution to the application and getting expanded systems capabilities (Stoica and Andrei 2007).

An improved **AH** solution to the real-world application involves a triple adaptation. The first is an adaptation to the needs of individual users – the design solution is not a fixed one, but a flexible one that covers a wide operational range. A multi-functional hardware is the result of an optimal design of just a function implementation, which is extensible by switching among different implementations. The second adaptation is to changing environment involving the environment, the customer requirements and the application targets that may change. As a consequence, the hardware adjustments are required for not an adaptation, but to an evolution as well. Adaptation to fabrication (manufacturing) imperfections is implemented by embedding adaptive characteristics into microelectronic hardware. It is implemented by a fine tunability of the components or by fault tolerant computing methodologies, with respect to the components imperfections or malicious induced catastrophes.

A hardware adaptation is looked as efficient if the systems capabilities are expanded by optimization /improvements of signal processing (adaptive filters, adaptive compression, adaptive /compressive sampling) or by optimizing the communication (bandwidth optimization, avoiding jamming and **EW**).

The configuration of an **AS** can be implemented by software or by hardware. The last one is not so popular as the first one. **AHS** provide crucial adaptive capability fitting some refined characteristics of adaptation: a nonstop system availability featured by no system down or interruption time; an adaptive reaction of an ultra-fast setting up time; an optimal functional synchronization at different level of **AHS** adaptation; an user –friendly **AHS** flexibility in setting and targeting new adaptation tasks in real time; a smooth adaptation with minimal overhead functionality.

By improvements in economics, both the cost per function and the cost barriers are constant envisaged so as the user technical benefits can't offset higher costs per a given technical function. Currently, the **AH** circuits tend to achieve the **ASIC**-level economic efficiency. As a consequence, the **AH** programmed **FPGA**, **FPAA**, **FPIC**, open-architecture **FPGA** began to displace the **ASIC**s in some areas as consumer devices and communication terminals. But still a focus of the **AHS** community must increase the efforts in reducing design cost per function and programming so as **AHW** implementations to be competitive indeed to equivalent **ASIC** implementations.

An engineering perspective of **EHW** terminology, design methods and application must rely on some practical and direct application related aspects that are of immediate help for the practitioners which are involved in any **EHW** area.

EHW means nothing else than *Reconfigurable Hardware* and a *Reconfiguration Mechanism*. In a narrow sense, **EHW** is a programmable hardware self-configurable by built-in Evolutionary Algorithms, a hardware that can change antennas, electronics equipments or circuitry, **MEMS** and **BioMEMS**. This is done by its *intelligent part*, the built-in mechanisms controlling the adaptation/self-configuration or changes either algorithmic, search-optimization relied or knowledge-based. The self-adaptation in most **EHW** systems is accomplished in order to survive unanticipated environmental changing conditions or even system failures that can put the system out of order definitely and forever. **EA**, but **GA** in main, are the techniques used most frequently to implement the search strategy for good reconfigurations implementing the system ability of being self-adaptive. The option is usually made in favor of **GA** because of **EA** general-purpose applicability: feasible implementation both in hardware and in software; no an accurate knowledge of the problem domain required; the optimal solution is not found by following typical general design rules.

Regarding the **EA** that are used by **EHW** practitioners, **GA** is the most popular, but some application use the ***Evolution Strategy*** (**ES**), namely predominant the (μ,λ) **ES** and not so much the $(\mu + \lambda)$ **ES**. The preferential use inside the **ES** frame is technically justified by the algorithmic behavior as follows: the (μ,λ) **ES** versions provide a huge exploration potential because its parents only live one generation; the $(\mu + \lambda)$ **ES** versions have a diminished potential so they are not so effective because parents can survive over many generations.

The **GA** used in **EHW** has a radically different task than a general purpose **GA** and it runs as a consequence. It is crucial for the practitioners to have a clear understanding of what features should a **GA** have to minimize the search time for **EHW** configurations and how this kind of **GA** behaves as a consequence of its specific aims. In **EHW** applications the **GA** is not asked for starting from a random initial population and to converge to an optimal solution (an initial design of hardware) in minimum time. This **GA** must face out unanticipated environmental condition, which suddenly occurred and alters the fitness landscape of the search space making the previous hardware configuration inadequate. This event is the start of a new search, but a search for a reconfiguration, which is totally distinct to the search for an initial design. Inside the framework of reconfiguration searches, the final population from the previous **GA** runs forms the initial population for the new **GA** run. A natural lack in genetic diversity is the consequence of that population convergence around the old optimum. Beginning with a localized initial population minimizes search time if the old and new optimums were near each other. But the **GA** still needs additional features that permit a widespread exploration of the search space: it must use reproduction operators capable of making large movements over the landscape for reducing time for reconfiguration. Also a high selection pressure may reduce the time for reconfiguration by using proportional selection, elitism or the truncation selection.

Another important **EHW** issue must be solved besides finding the particular strategy of adaptation in order to keep the functionality (Garrison et al 2003): a suitable response time for the adaptability - not just only the suitable adaptability time required by that application hardware, but the not desired consequences if this time threshold was not met. Usually, the **EA**-programmed **FPGA** are featured by outstanding execution times if the designer takes care about the fact that designing an **EA**-programmed **FPGA** is a time-consuming process; compulsory, the **EA** parameters (the selection methods, the reproduction operators) must be chosen before the process of **FPGA** implementation effectively starts.

Dramatic changes are happening in the relation between hardware and the application environment, in case of malicious *faults* or need for *emergent new functions* that claim for in-situ synthesis of a totally new hardware configuration.

EHW is suitable for flexibility and survivability of autonomous systems as that ones developed by **NASA JPL**.

EHW *survivability* means to maintain functionality coping with changes in hardware characteristics under the circumstances of adverse environmental conditions as for example: temperature variations, radiation impacts, aging and malfunctions.

EHW *flexibility* means the availability to create new functionality required by changes in requirements or environment.

The technical objectives of most complex **EHW** relied systems involve development of flexible and survivable hardware capable of intelligent self-configuration, self-tuning, and self-repair, that can adaptively change through reconfiguration/compensation for optimal signal processing, sensing/control, and survival in the presence of faults/degradation. Successful applications are developed in a large range of fields as: automated design, automated calibration and tuning, in-field adaptation of hardware systems, sensing, control and robotics.

The application developer may meet different design tasks to be evolved. As the case the design to be evolved could be: a program, a model of hardware or the hardware itself. Algorithms that run outside the reconfigurable hardware mainly feature the actual **EHW** state of the art, but also some chip level attempts were done. Future solutions will be the Integrated System at a Chip and IP level.

The path from chromosome to behavior data-file is different in case of intrinsic and *extrinsic* **EHW**; see this difference expressed by two simplified block diagrams as in Fig. 2.1, (Negoita and Stoica 2004).

A quantitative comparison of Evolution in Simulations versus Evolution in Hardware has a higher weight from an engineering perspective at a first glance. **EHW** Evolution in Simulation has some typical features and parameters value (Negoita and Stoica 2004) as follows :

- Computationally intensive (640,000 individuals for about 1000 generations)
- Runs over tens of hours, expected about 3 min in 2010 on desktop PC for experiments with netlists of about 50 nodes
- SPICE scales badly, namely time increases nonlinearly with as a function of nodes in netlist, in about a subquadratic to quadratic way
- No existing hardware resources allow porting the technique to evolution directly in hardware (and not sure will work in hardware)

Regarding the evolution in hardware the following quantitative details are of practical interest:

- JPL's VLSI chips allow evolution 4+ orders of magnitude faster than SPICE simulations on Pentium II 300 Pro.
- About tens of seconds were got for circuits of high complexity as from Koza's experiments

A relevant qualitative point of view emphasizes the huge advantage of the on-chip **EHW** versus CAD/synthesis tools.

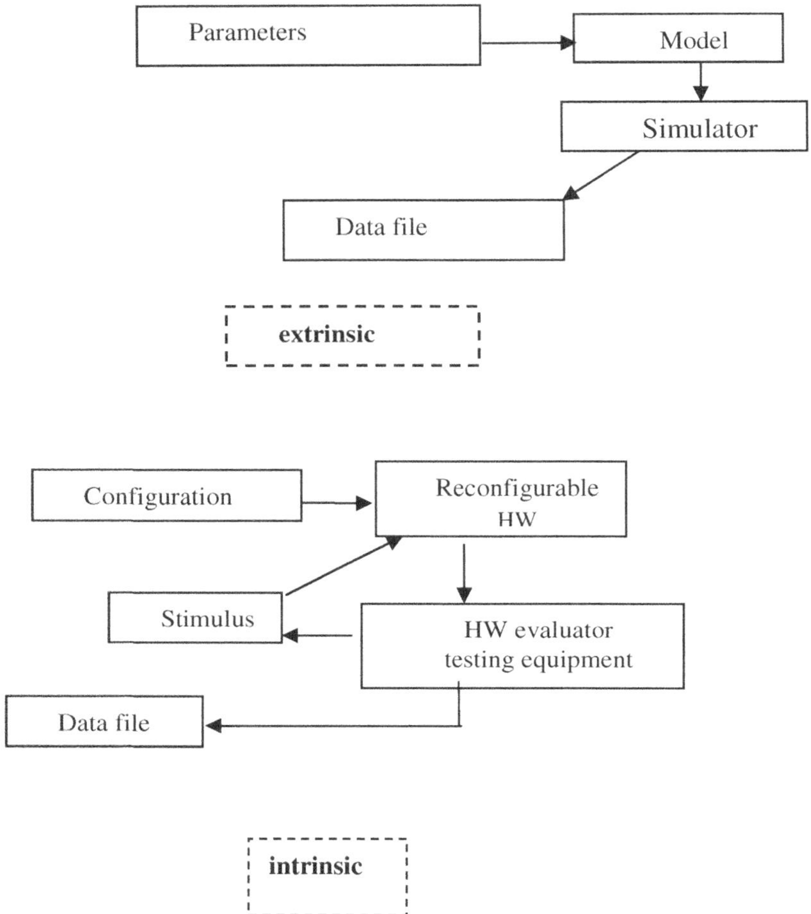

Fig. 2.1 Extrinsic and intrinsic **EHW -** a simplified Block Diagram comparative

EHW can overcome a lot of technological manufacturing problems of the electronics integrated circuits: fabrication mismatches, drifts, temperature and other plagues to analog, exploiting the actual on-chip resources – finding a new circuit solution to the requirements with given constrains and actual on-chip resources.

It is suitable to briefly underline what a multi-stage search regarding **EHW** evolution practically means from an engineering point of view.

EHW multi-search deals with a search for the circuit topology, followed by the optimization of circuits' parameters. It runs over two main distinct design stages:

- **Stage 1:** the *GA-based Evolution of the topology* of the electronic, integrated circuit

- **Stage 2 :** *GA- based Optimization* of the electronic devices ,*the transistors sizes* mainly, for the best circuit topology resulted in Stage 1. The initialization is performed in Stage 2 with the best circuit topology and random parameters

Fig. 2.2 presents a multiplier that was evolved by NASA JPL EHW Group through a multistage research consisting of 200 generations during the first stage and 40 generations during the second stage

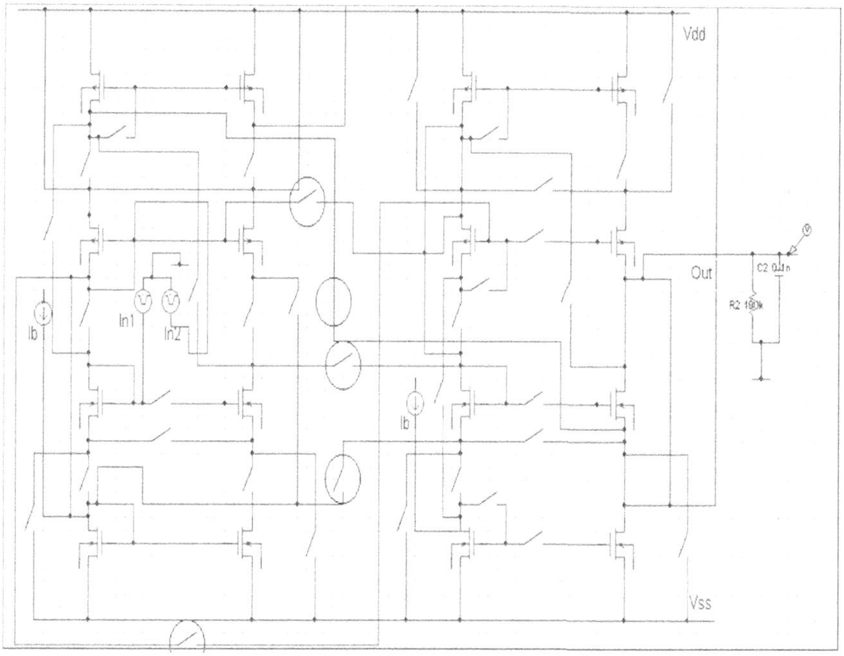

Fig. 2.2 A multiplier evolved through multi-stage research, after NASA JPL EHW Group

2.1.2 EHW Technological Support (FPGA, FPAA, FPTA, FPMA, PsoC)

The idea of applying evolution to artificial systems has surfaced thirty years ago, but the technology available at the time was not proper to implement this methodology in hardware.

The development of both computing technique with increasing computational power and the appearance of programmable integrated circuits, especially their new generation – **f***ield* **p***rogrammable* **g***ate* **a***rrays* (**FPGA**s) and most recently **r***e-configurable* **a***nalogue* **a***rrays* (**FPAA**s) and **f***ield*-**p***rogrammable* **i***nterconnection* **c***ircuits* (**FPIC**s) or configurable digital chips at the functional block level, (*open-architecture* **FPGA**s) make possible for most companies to evolve circuits as the case would be.

Good times are now for electronics engineers, this profession is deeply changed, evolutionary approach applied by **FPGA**s or open-architecture **FPGA**s to electronics make hardware architectures as malleable as software, evolution don't care about complexity as the evolving system works, the only limits to electronics design are the limits imposed by our own understanding.

2.1.2.1 Introduction to Programmable Integrated Circuits

Integrated circuits are called *programmable* when the user can configure its function by programming (Sanchez 1996). This type of circuit is delivered after manufacturing in a generic state and the user can adapt it by programming the particular function on request.

In this sub-chapter we will tackle only notion concerning *programmable logic circuits* – that means the programmable function of these circuits is a logic one, any is their type, from simplest Boolean functions to sophisticated state machines.

A programmable logic circuit is an array of elementary devices: functionally complete devices or universal function. A functionally complete device (*NAND* or *NOR* gates, for example) is possible to realize any logic function as an interconnection of several devices of the same (their) type. A device capable of realizing by itself every logic function of *n* variables (the multiplexer, the demultiplexer and the memory, for example) is called a *universal function*.

The programmable logic circuits can classify concerning the type of programming and the internal organization type.

The programmable circuits may have a type of programming either *irreversible* – by burnt fuses, with impossible further correction of the program – or *reprogrammable* – by programmable internal functions and interconnections by a program controlled with memory cells; the program can be erased by *UV* light or electrically as the type of circuit is.

The internal organization of a programmable logic circuit leads to three classes as will be described below.

1. ***P**rogrammable **l**ogic **d**evices* (**PLD**) – an array of **AND** gates receiving the system logic inputs and generating product terms, these being followed by an array of **OR** gates which generate the logic outputs of the system (see Fig.2. 3).

There are known three **PLD** variants:

- **PROM** (**p**rogrammable **r**ead-**o**nly **m**emory) – only the *OR* gates array is programmable, that means one can select the fixed input *AND* gates to connect to each *OR* gate
- **PAL** (**p**rogrammable **a**rray **l**ogic) – only the *AND* gates are programmable, that means the input configuration of *AND* gates is programmable, but the input configuration of *OR* gates is fixed
- **PLA** (**p**rogrammable **l**ogic **a**rray) – both *AND* and *OR* gates arrays are programmable

The **PLD**s drawback is their fixed character of interconnections – that means it is possible to program the function but not the interconnection between functions in case of trying to realize a multilevel function.

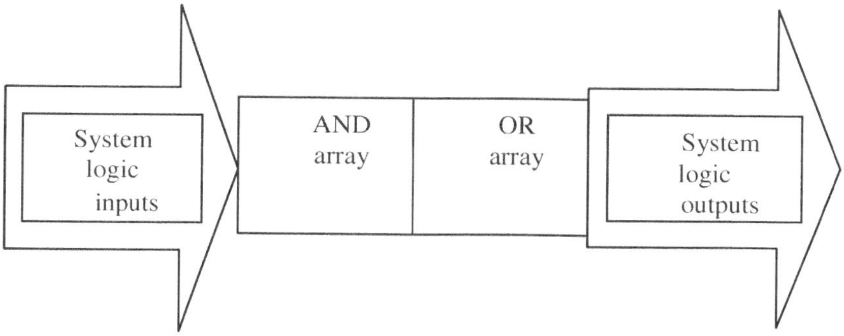

Fig. 2.3 PLD simplified block scheme

2. C*omplex* p*rogrammable* c*ircuits (***CPLD***).*

A number of programmable cells realizing a universal *n*-variable logic function have their inputs electrically connected by an interconnection network from the system external logic inputs and/or the outputs of themselves.

The interconnection programmability is limited, that means only one possible connection is possible between two points, and that because the interconnections are pre-routed in advance by the manufacturer, the user only choosing of these interconnections. This disadvantage over the **FPGA** is blurred by the main advantage of completely time predictability of establishing the circuit configuration.

3. **FPGA** *circuits.*

The engines of impressive **EHW** development nowadays are **FPGA**s. A **FPGA** is defined as an array of logic cells (see Fig. 2.4) – both functionally complete devices and/or universal functions – placed in an infrastructure of interconnections which can be *programmed* at the following levels to realize a certain function:

- the *function* of the logic cells
- the *interconnections* between cells, with different types of interconnection and several path possibilities between two points in the circuit – this is a major advantage given **CPLD**s, despite of the timing unpredictability before the final routing of the circuit
- the *input/output cells* are programmable for the direction of information, storage element, and electrical level.

A comparison between **FPGA**s and gate arrays leads to the following remarks:

- the complexity is comparable
- the time of circuit realization: the working state of a **FPGA** starts a few minutes after the on spot programming step, given to some month for the gate arrays

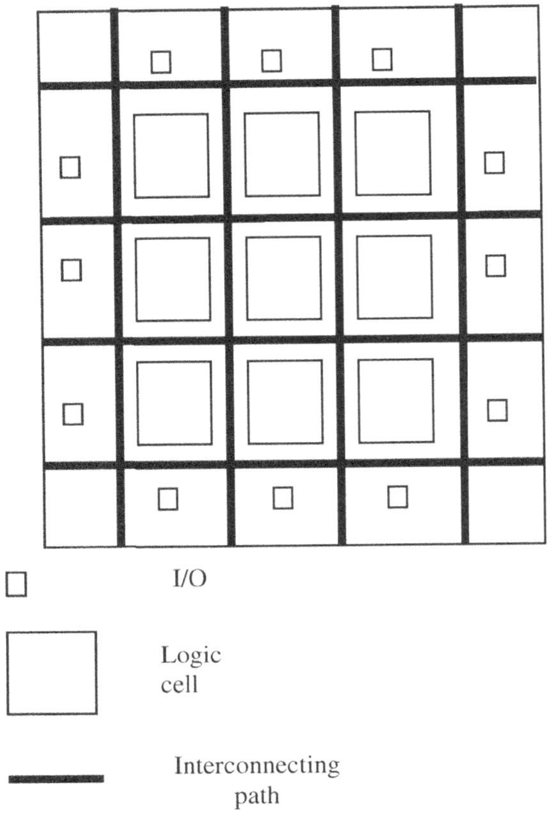

Fig. 2.4 **FPGA** simplified block scheme

- the market point of view: the production cost of **FPGA** is preferable for small series – a few of thousands - of pieces, but a better cost is got for gate arrays in case of series consisting in more than ten thousands of pieces.

2.1.2.2 FPGA Families and Advanced Type of FPGA

The complexity level of the architecture in a logic cell (Sanchez 1996) leads to a **FPGA** classification in two families:

- *fine-grained circuits* - have several functionally complete devices or some universal function in its structure; optimal cell utilization is allowed despite of the difficulty in routing the circuit
- *coarse-grained circuits* relied on basic logic cells in form of complex universal function, with several logic output variables; an easy is allowed routing but with a less optimal utilization of the integrated circuit surface.

Another classification criterion – the programming technology - also divides **FPGA**s in two families :

- **SRAM** *technology based* **FPGA** – use a **RAM** reprogrammable technology, namely each programmable point, either in interconnection or in function region, is controlled by a bit belonging to a static **RAM**
- **antifuse** *technology* - uses a non-programmable technique, namely the **antifuse** – a point with very great impedance (in case of open connection) is transformed in one with very weak impedance (in the situation of closed connection).

Antifuse technology has still its own advantage over the **SRAM** technology by using silicon chip surface that is very small given those used for **SRAM**. **FPGA** are suitable for implementation of systems with multilevel logic because both the structures of their logic blocks and the interconnection possibility allow it. Much more on design methodology and the steps of the design process with **FPGA** is to be seen in (Jenkins 1994), (Xilinx 1995) and (Actel 1995).

It is beyond the main purpose of this book to deeply describe the main features of **FPGA** families delivered by different manufacturers, namely their blocks architecture, structure of interconnection and the programming manner. We only limit ourselves to remind that one of the world leaders among the **FPGA** market suppliers , Xilinx, produces now the family 6200 – *open architecture* **FPGA***s* – the most suitable family for complex **EHW**. No details regarding the transparent cellular architecture of Xilinx 6200 chip, either the programming bit stings or the configuration switches and the hierarchy of routing resources will be introduced. Just the the Xilinx XC6200 cell - a **SRAM**-controlled switch(mux) – based FPGA – is illustrated in Fig 2.5.

Let's therefore to give some general details about the advanced **FPGA** family from Xilinx, because they seem to be the further engine of **EHW** technological support, leading to outstanding applications such as virtual computing.

The *virtual computer* is a reconfigurable hardware system, allowing a custom processor chip for one special application, by request. This *reconfigurable hardware system* (also called *reconfigurable computing system*) is produced in form of a PC or Workstation plug-in board, acting as a co-processor to the **M***ain* **P***rocessing* **U***nit* (**MPU**). The main application program contains a special sub-routine that downloads a digital chip design into the **R***econfigurable* **P***rocessing* **U***nit* (**RPU**) of the Reconfigurable Computer, some calculation normally achieved by **MPU** being made in this situation by **RPU** instead.

This method of implementing some calculations directly in a custom devoted computing chip are advantageous because of two motivations: first they are the fastest calculation manner for any type of specific calculus and second, the computer power is increased by this flexibility in hardware, optimising the configuration of **RPU** as dictated by the application necessity.

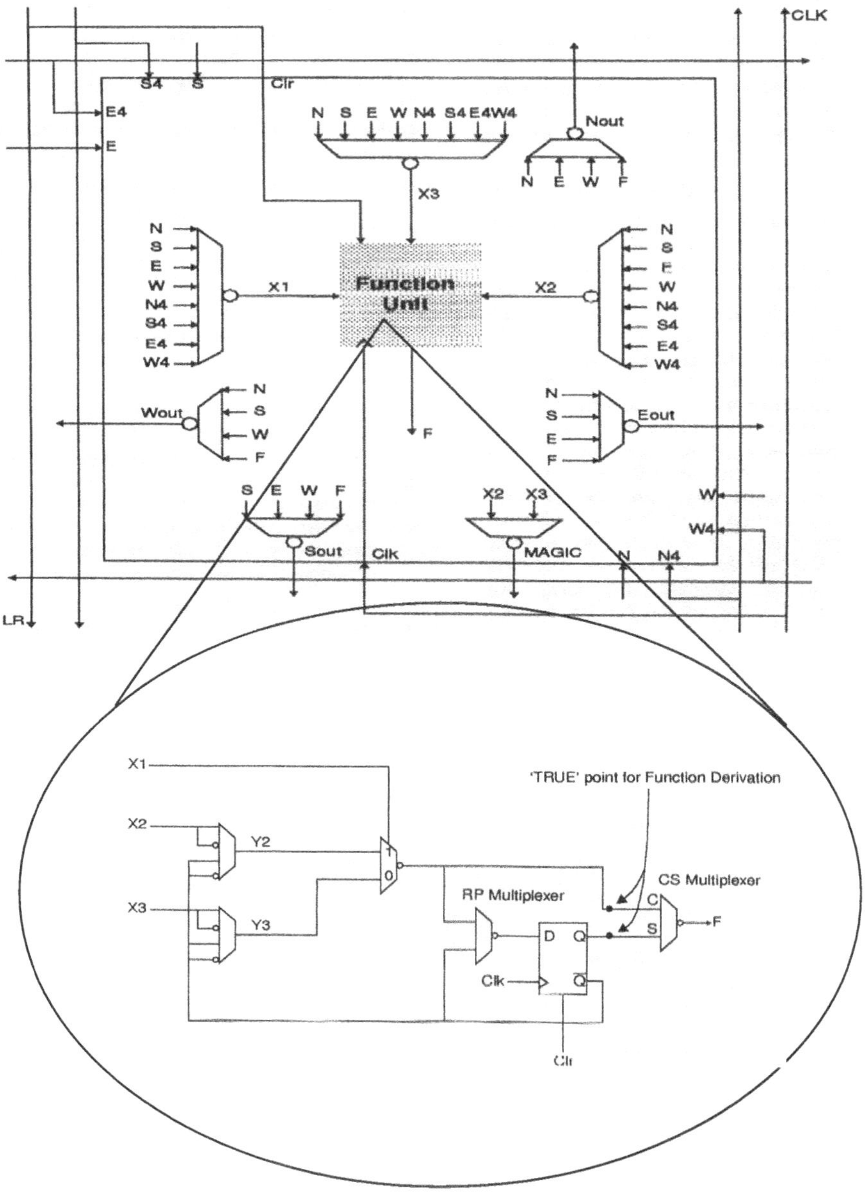

Fig. 2.5 The basic Xilinx XC6200 cell

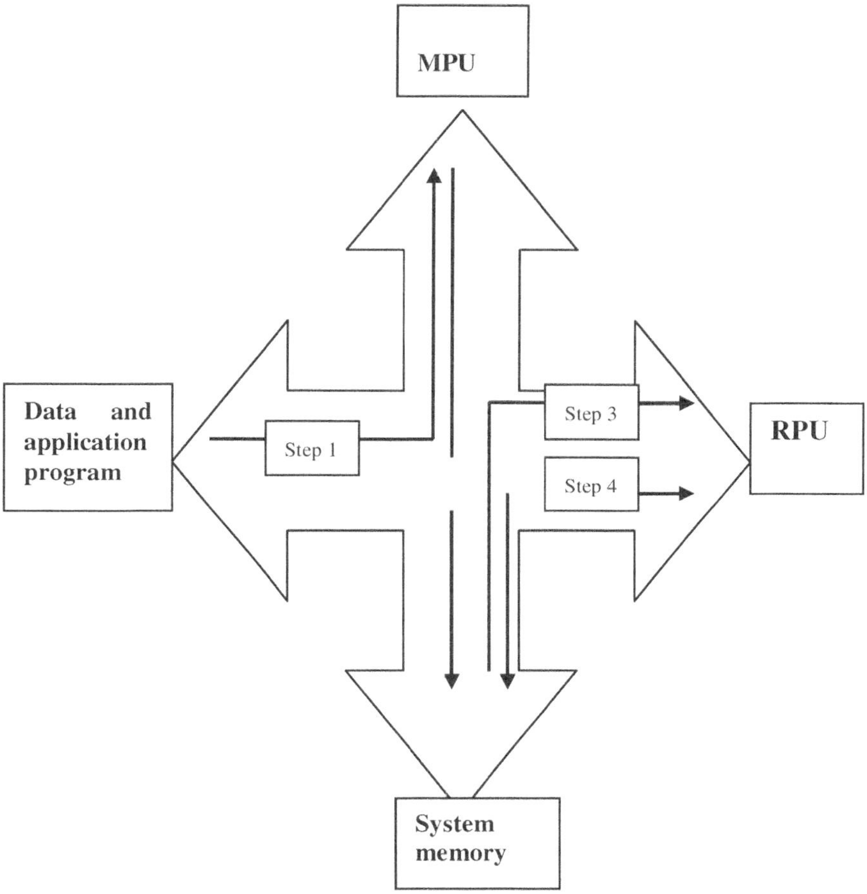

Fig. 2.6 The simplified block scheme of a reconfigurable hardware system after (Virtual Computing Company 1997)

The simplified block scheme of the reconfigurable computer and of application running is depicted in Fig.2.6, after (Virtual Computing Company 1997). The main blocks of this architecture are :

- the **M***icro***P***rocessor* **U***nit (***MPU***)* – managing the computer functions and running the application program
- the *memory* – storing the application program and the data to be used
- the usual *I/O devices* for displaying to display or store data and programs
- the **RPU** allows its own reconfiguration by software and *perform one unique calculation* the fastest possibly; after this unique calculation is performed independently of **MPU**, the result is returned to **MPU** and **RPU** is prepared to be reconfigured again for another unique calculation.

The steps of running the application program by a reconfigurable computer system are as follows:

STEP 1 – program and data getting by **MPU**
STEP 2 – program and data saving to **memory**
STEP 3 – **MPU** downloads the program for the unique calculation in **RPU**
STEP 4 – **RPU** makes multiple readings or writings of data to be processed by the main application program back and forth to the system memory; after the unique calculation allowed inside the **RPU**, the **MPU** is prepared to perform another calculation.

Final concluding remarks on reconfigurable computer architecture hardware:

- no clock cycles are used by **RPU** for general function routines as by **MPU**, only the unique calculation is achieved
- open-architecture **FPGA** has a reprogrammability of **SRAM** type.
- the main **RPU** blocks are: configurable logic blocks, programmable interconnected resources, I/O blocks and blank logic

The Virtual Computing Company delivers *H.O.T. Works 6200* Development System, a complete programming and development system for *Xilinx 6200* **RPU**. It includes the *PCI- XC6200* board – a single PCI Bus IBM compatible plug-in board for real-time emulation allowing the possibility to explore the behaviour and attributes of the *XC6200* family. Attached is a software package too: the *Lola* Programming System (both an object oriented high level hardware description language and its compiler), including the *XC* Editor (a graphical layout editor with a circuit checker, mapper, placer and router) and the bit stream generator/loader for the *XC6200* family.

Attached are other necessary software tools, for example a Hardware Object Technology Interface for insertion of designs into executable C language programs, see also (Virtual Computing Company 1997) for more details.

2.1.2.3 Field Programmable Analog Arrays(FPAA)

The appearance of **r***econfigurable* **a***nalogue* **a***rrays* (**FPAA**s) was crucial for the technological support required by companies involved in electronics research and development as well as in manufacturing. The analog reconfigurable hardware allows prevention or removal of essential fabrication mismatches and other refined technological problems by evolving circuits as the case. Analog reconfigurable hardware has actually a huge weight in **EHW** environment, see Fig.2.7, where the actual hardware platforms for **EHW** development and their authors are enumerated (Negoita and Stoica 2004).

FPAA reconfigurability relies on switched capacitors, typical to the main providers as Pilkington, Motorola – see Motorola MPAA020 – or Anadigm (see Fig. 2.8).

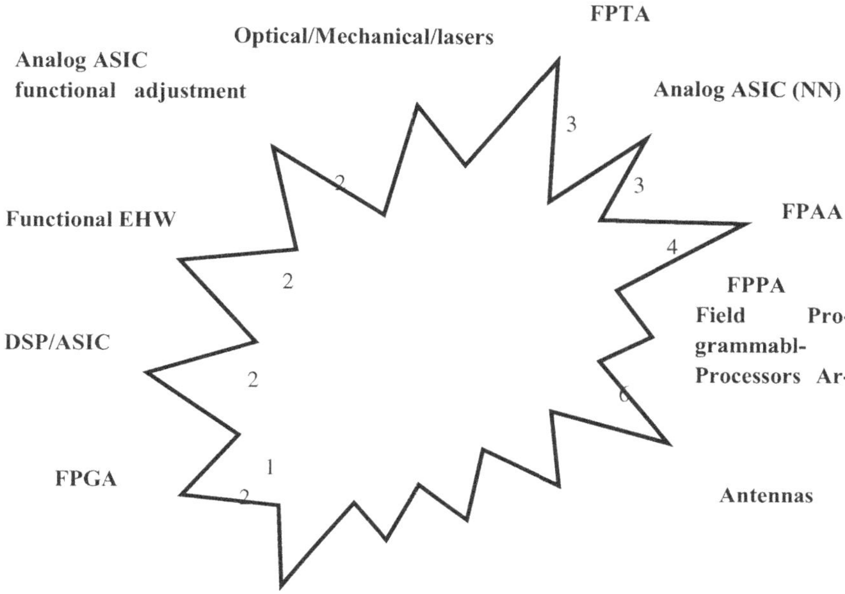

1 Thompson U, Sussex, UK
2 Higuchi T, ETL, Japan
3 Stoica A, NASA JPL
4 Zebulum,R NASA JPL
5 Marchal P, CSEM, Switzerland
6 Linden

Fig. 2.7 The main hardware platforms for **EHW** development and their authors/providers

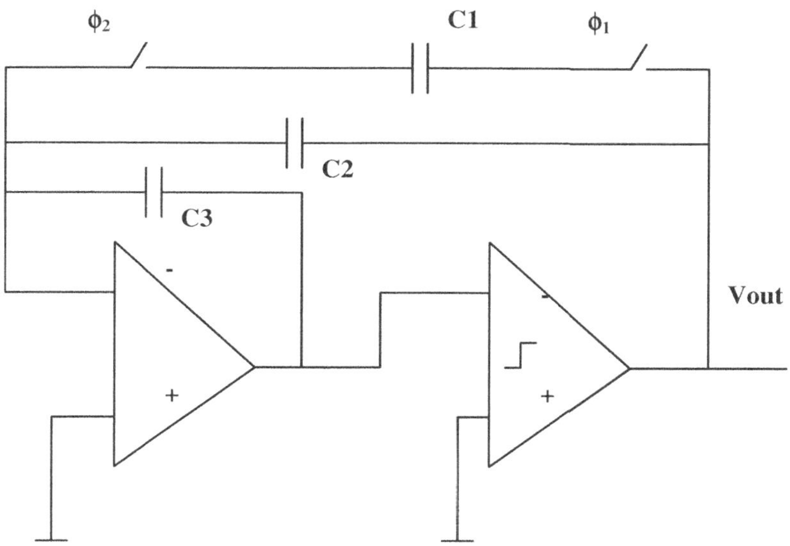

Fig. 2.8. A typical **FPAA** switched capacitor configuration

Zetex **TRAC** provides a totally ***R**econfigurable **A**nalog **C**ircuit* structured in form of 20 cells, each being an op-amp with a small reconfigurable network. Such a cell performs any of the following analog operations: add, negate, subtract, multiply, pass, log, antilog, rectify, or basic inverting op-amp for use with external components.

Lattice ispPAC10 - see Fig. 2.9- contains four programmable analog modules and a programmable system. This circuit is an **EHW** analog reconfigurable hardware that can be configured to implement 2nd and 4th order active LP and BP filters in the 10khZ – 100 KHz range.

High complex architectures are featuring some analog reconfigurable hardware of the Anadigm **FPAA** family, as for example AN220E04.

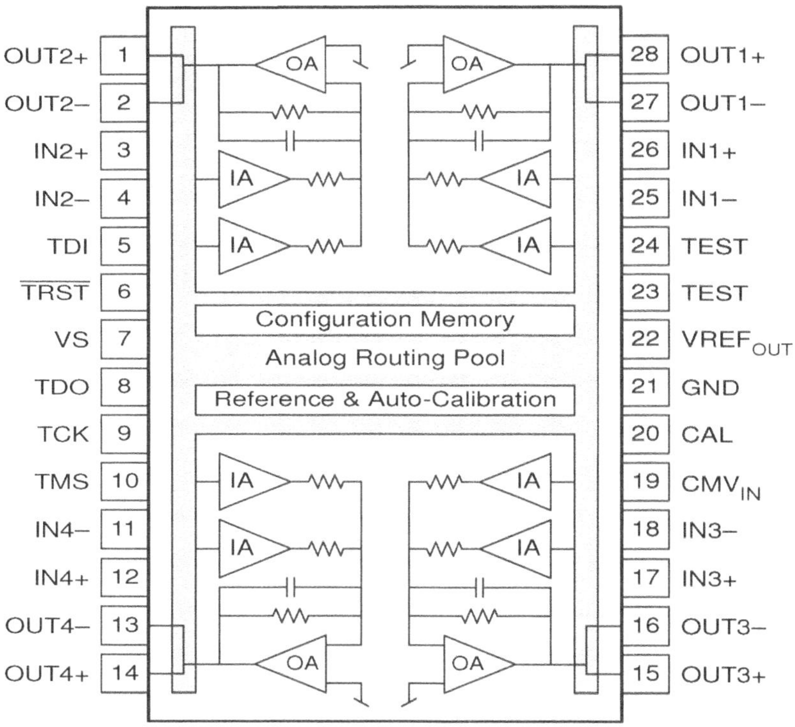

Fig. 2.9 The block diagram of Lattice ispPAC10

2.1.2.4 Field Programmable Transistor Arrays (FPTA) Produced by NASA JPL

The programmability of **FPAA**s is limited to only allow configuration around op-amp level. But the applications require also for many interesting circuit topologies to be evolved below the op-amp level. This application requirement led to another kind of programmable (reconfigurable) hardware relied on *evolution-oriented devices* that have some advantages over **FPAA**, namely: can reprogram many times, can understand what's inside and are featured by a flexible programmability. This is the so-called *custom made* **EHW**-*oriented reconfigurable hardware*. Some briefly comments are to be made regarding the types of *custom made* **EHW**-*oriented reconfigurable hardware* (Negoita and Stoica 2004) :

- **JPL PTA**s are chips of reconfigurable hardware at transistor level, both analog and digital
- **FPTA chip of Heidelberg University** is an array of 16 x 16 transistors, performing a programmability in connectivity and channel length
- **JPL'98 FPTA-0** chip is a programmable transistor array cell with 24 programmable switches, a sufficient number for meaningful circuit topologies. All three terminals of a transistor cell are connected via switches to expansion terminals. The chromosomes give the value HIGH-LOW of the switches (not only ON-OFF).
- **JPL'2001 FPTA-2** chip is a second generation reconfigurable array chip, a programmable array of transistor array cells implementing an evolution-oriented reconfigurable architecture, featured by a NESW interconnection amongst 64 integrated cells (an 8x8 matrix of reconfigurable cells), each of the cells having 44 transistors. It is the first chip integrating reconfigurable processing circuitry with sensing: an array of 16x8 photodetectors distiributed within the cells is also integrated on this chip. **FPTA-2** chip is able of receiving 96 analog/digital inputs and provides 64 analog/digital outputs, it is the first **FPMA** (**F***ield Programmable* **M***ixed-signal* **A***rray*)
- **PAMA** (**P***rogrammable* **A***nalog* **M***ultiplexer* **A***rray*) chip is an analog platform based on analog multiplexers/demultiplexers (Santini , Zebulum et al 2001). The multiplexers/demultiplexers are fixed elements that perform the interconnections of the different discrete electronic devices that can be plugged into the board. The platform performs intrinsic evolution of analog circuits through a **RM** – represented by a **GA**. Each gene configures the select input signals of a particular analog multiplexer. A multifunction I/O board is connected to the PC bus to perform the A/D conversion and the chromosome download. The control bit strings (**GA** chromosomes) are downloaded to the **RH**. The circuit evaluation runs as follows: circuit responses are compared against specifications of a desired response, this comparison being followed by a circuits ranking based on how close they come to satisfying the target.

  PAMA provides a practical environment to evolve generic analog circuits based on discrete components, without the need of simulators. This is in fact a very useful prototype platform allowing a large number of component terminals and evolution of a great number of circuits. The circuit evaluation speed is featured by 6 minutes to evolve a certain circuit.

  Other strong features is that **PAMA** platform provides protection against illegal configuration that may damage electronic components and confers the possibility to analyse circuits which have been evolved , due to access to individual circuit elements with test equipment.

An innovative powerful **P***rogrammable* **S***ystem*-**on**-**C***hip* (**PSoC**) is the family PSoC™ CY8C25122/CY8C26233/CY8C26443/CY8C26643 of configurable mixed-signal arrays provided by Cypress Connects.

PSoC is a chip, reconfigurable device that replaces the components of a multiple **MCU**-based system components. A chip of this kind includes configurable analog and digital peripheral blocks, a fast **CPU**, Flash program memory and **SRAM** data memory in a range of convenient pin-outs and memory sizes.

2.1.2.5 Practical Remarks on the Technological Support of EHW

Reconfigurable hardware performs function changes by configuration changes. These are switch-based devices that change cell function and cell interconnections, switches interconnecting functional modules of primitive analog and or digital functions.

Vendor programming tools allow switches to be turned ON/OFF, in a mode that is visible or not invisible to the user, via intermediary program conversions. To set up the status of the switches, namely which switches are ON and which are OFF, is the search/optimization problem for **EHW**. Status of switches – ON or OFF – can be straightforward associated with a binary representation used by **GA**.

Different strategies may be applied to set up the switches when unidentified faults prevent mapping of computed solutions: a local search for compensation; variations around a configuration determined by knowledge/analytical means or a new configuration needs to be searched.

The language for programming reconfigurable hardware must define:

- an *alphabet* ,expressing choices of cells and
- a *vocabulary/grammar*, expressing the rules of interconnect.

An evolvable hardware system, any is its destination, either for demonstrations, prototype experiments or real time implementation, must be structured in form of two main components:

- the **R***econfigurable* **H***ardware* (**RH**) and
- the **R***econfigurable* **M***echanism* (**RM**) [(Stoica, Zebulum et al 2002)

Regarding the practical ways of implementing an evolvable system, real-world applications requirements are toward a reliable solution featured by compactness, low-power consumption and autonomy.

A solution that proved to be effective in various applications is the evolution on the **JPL SABLES** (**S**tand-**A**lone **B**oard-**L**evel **E**volvable **S**ystem), as from (Zebulum, Keymuelen et al 2003).

The main integrated components by the **JPL SABLES** are as from Fig. 2.10:

- the transistor level **RH**, namely a **JPL FPTA**
- the **RM**, consisting of the evolutionary algorithm, implemented by a **TI DSP** acting as a controller for reconfiguration.

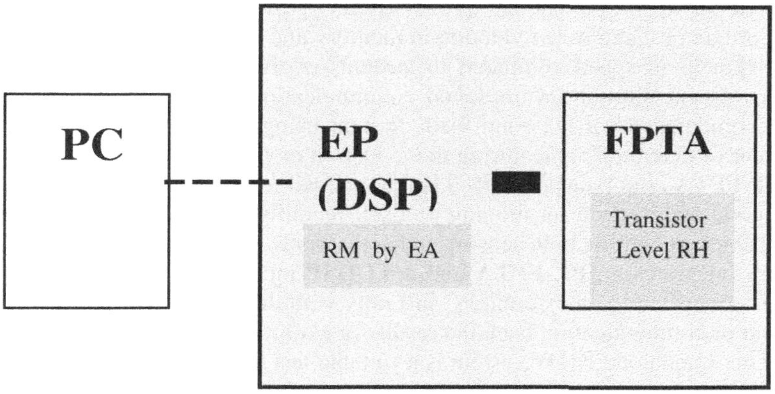

Fig. 2.10 The simplified bloc diagram of **JPL SABLES** (after Zebulum, Keymuelen et al 2003)

The information flow and implementation, see Fig. 2.11 with **TI DSP/FPTA** chip pair of **JPL SABLES** is featured both by autonomy and speed in providing on-chip circuit reconfiguration: about 1000 circuit evaluations are performed per second.

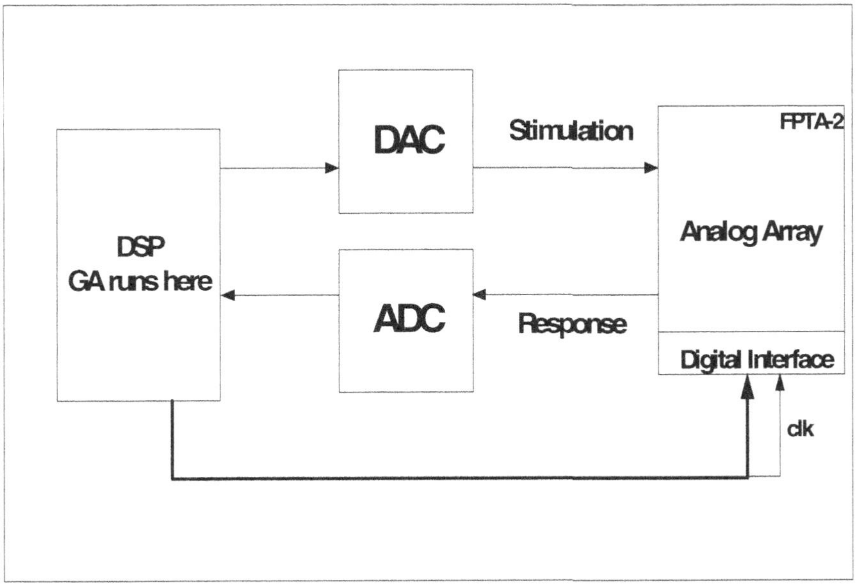

Fig. 2.11 The information flow between **DSP** and **FPTA** in **JPL SABLE**

Another detail of the parameters **JPL SABLES** performance are the followings: 1 – 2 orders of magnitude reduction in memory and about 4 orders of magnitude improvement in speed compared to systems evolving in simulations. The speed improvement through an improved communication is of about 1 order of magnitude compared to a PC controlled system using the same **FPTA** whilst about 1 order of magnitude is featuring the reduction in volume.

The **DSP/FPTA** chip pair performs a fast download for evaluation of individuals. It is a good architecture for moving to a self-reconfigurable system-on-a-chip that is fault-tolerant and fits both sensors and actuators work environments.

SABLES integrates a **JPL FPTA** and a **TI DSP** into a system that is standalone and is connected to an external PC just only with the aim of receiving specifications and of communication back the results of evolution at different stages for analysis. This standalone **EHW** system is a suitable test bed for the evolution and recovery of electronic circuits by applying faults to entire **FPTA** cells.

SABLES is a practical tool for design and development of electronic circuits and equipment that are intrinsic insensitive to faults by using on board **EHW** to achieve fault tolerant and high reliable systems. These systems and components are required for applications not only in aerospace area, but in any sharp environments featured by extreme temperatures and dangerous level of radiation.

2.1.3 EC Based Methods in EHW Implementation: EHW Architectures

This sub-chapter is an overview of some present methods that the scientists breaking new ground in the area of **EC** use to implement these techniques and systems adaptation in hardware. The first part presents the main features of behavioural artificial intelligence; the second part is a condensed presentation of the main **EHW** architectures types. Some examples of **EHW** systems are applied in well known nowadays areas such as analogue and digital electronic circuits, cellular machines, controllers for autonomous mobile robots, pattern recognition and special neural networks, namely with dynamic topologies ones, and this happens despite of the fact that **EHW** implementation is in a pioneer stage.

Biological models are applied in computer technique, adaptation and real-time learning as a natural consequence of moving from classical (hard) computing to **S***oft* **C***omputing* (**SC**). This allowed the passing from classical (*knowledge based* **AI***)* to *behavioural computational intelligence* as we mentioned in the previous parts of the book too. The main typical features of this new type of **AI** are as follows (Negoita 1995):

- any behavioural function is *collectively generate* by elementary component functions
- behaviour is the result of interaction in *local (elementary) components,* no more centrally controlled as in classical **AI**
- a design from *peripheral elements to the central parts* is used for these systems

- information processing has an outstanding fluidity achieved by *complementary transformations* as:
    - emergence/disappearance
    - increasing/diminishing
    - combination/separation
    - autonomous change, for example metabolism or natural selection.

The common framework of *behavioural* **AI** and of **SC** made possible the implementation of hardware circuitry with intrinsic logic elements specific to **EC**. This was called by real-world applications as those typical in environments where human capacity of intervention is limited (nuclear plants, space applications, etc...) as from (Shimonara 1994), (Higuchi et al 1994), (Negoita and Zimmermann 1995) and (Negoita 1996).

Among the fundamental features of a living being two are most essential for computer technique: *parallelism* and *co-operation of parts.* These are in fact the main features of the architectures used in hardware implementation of **EC** elements. In this case the parallelism is "massive grain" that means a deeply developed one at different levels of the architecture.

The biological metaphors used to implement hardware architectures with intrinsic **EC** elements use two modelling principles:

- *life-like modelling*, that means changes based on embryological principles and
- *social-modelling,* that means a dynamic process in which macroscopic state of systems influences microscopic components and vice-versa (Negoita 1995).

The final aim of **EC** techniques development and of their silicon implementation is to create architectures towards an artificial brain, a computer having its own ability of reasoning or decision making, but also being able of emergent functionality creation, or having the possibility of self-creation and evaluation of its own structure.

RH consists of some chips composed of small generic circuits that are programmable both as functionality and as interconnection. Automatic tools perform programming and configuration of the **RH** chips. These tools translate the high-level description into a circuit-level description. The circuit-level description can be mapped and routed on a particular architecture of the chip that is to be reconfigured.

FPGA are traditionally most frequently used reconfigurable chips for small-series production especially, because they are able of providing fast and low cost way of implementing reconfigurable **AHS**. The advanced families of **FPGA** allow a more powerful integration and functionality options. This is the consequence of their circuit topology including specialized multiplexers, memory arrays and other interconnection resources.

The bunch of **RH** implementation methods enlarges and progresses very fast in the framework of the **HIS**. On its turn, **EHW** plays a crucial role in the implementation of **CI** technologies. The hardware implementation of some circuits with intrinsic non-conventional logic methods proves as very effective, despite being a nascent research area.

A suitable area for this strategy is the recent research progressing implementation of **I**_nterval_ **T**_ype-2_ **F**_uzzy_ **P**_rocessors_ (**IT-2 FP**). The requirement of new strategies for making faster type-2 fuzzy inferences are justified by being supportive to a set of real-time applications, especially those related to consumer electronics. The **IT2-FLS** implementation is relied on the hardware architecture of **IT-2 FP**. The proposed **IT-2 FP** architecture is more complex than other **FS** architectures. A hardware architecture intended to implement **IT2-FLS** in hardware by an **IT-2 FP** is proposed in (Melgarejo and Pena-Reyes 2007). There are four stages in this architecture:

- a fuzzification array
- a rule unit array that computes inference engine
- a type-reduction circuit that uses Wu-Mendel Bounds
  (both boundary centroids and uncertainty bounds are computed as non-linear functions; the computational complexity must be handled carefully for getting an optimal hardware implementation)
- an additional control unit that is in charge of controlling the data flow

The first two stages are organized as hardware arrays, the same functional block being instantiated many times in this stage. The computation of uncertainty bounds and the defuzzification are tasks of a pipelined arithmetic circuit. A special control unit supports the operation of the processor.

Dynamic reconfiguration properties of advanced **RH** can be capitalized to deal with real-world application requiring fuzzy adaptive processors. This can be made mainly at the level of rule unit array and rule computation units: if the rule base must be adaptive, the inference engine is to be computed by means of a reconfigurable unit array (Melagrejo et al 2004).

The hardware implementation of any **CI** algorithm is challenging, because of the finite computational resources supplied by the hardware context. Much more, some constraints must be satisfied with regard to operation frequency, implementation area, power consumption and thermal dissipation, for example. These reasons lead to a very practical engineering conclusion: the **EHW** developer must begin by considering a suitable computational model *before* a to establish a hardware architecture to the algorithm.. By such a design strategy, both the available resources are optimally handled and the required design goals by the application are achieved.

An overview on the most known types of hardware architectures with intrinsic **EC** logic elements is helpful for both the new comers and the traditional **AHS** and **EHW** practitioners. These architectures are the followings:

- *embryological* architectures
- *emergent functionality* architectures

- *evolvable fault tolerant* systems
- *parallel evolvable* architectures *(of Higuchi type)*

Embryological Architectures. The specialized **FPGA** for embryological architectures is in essence a lattice substrate of *multiplexer cells* with associated **RAM**. This architecture is suitable for hardware implementation of both combinatorial and sequential complex logic circuits. Circuits are grown in silicon by architectures that are *RAM-reprogrammable* **FPGA** – a two-dimensional array, which consists of dedicated (*specialized cells*) that can be interconnected by general interconnection resources. Cell instantaneous *self-reconfiguration* is allowed to each specialized cell because it stores the complete functional description of the whole circuit to be designed (evolved).

A specialized cell in its turn is structured as an array of proto-cells (four different types of *specialized type of sub-cell* achieving a given internal process). A proto-cell interacts both with other sub-cell of the same type in other cells and with ones of different type in the same cell.

Any specialized cell has a structure derived from the **B***inary* **D***ecision* **D***iagram* *(***BDD***)* tile and is suitable to directly implement this tile. During the second **EHW** design step (see sub-chapter 2.1), the designer must encode the available information about the circuit represented by the set of corresponding *minimized* **BDD***'s*. This was got *from the* **BDT** (**B**inary **D**ecision **T**ree) circuit description by regrouping identical sub-trees (in the same manner as the subroutines in algorithms).

The (encoding) programming field of a gene has two components: the *behavioural gene* (the f-gene) and the *topological (address) gene* (the s-gene).

The programming field of the whole circuit is called the *global genome, after* (Marchal et al 1996). This description of the whole digital organism (circuit) is stored in a genome memory located in each cell, namely in its nucleus-like proto-cell.

An embryological architecture has three-elaboration working steps (Marchal et al 1996):

1. Specialization - during which the coordinates of the mother cell are provided to the whole system (all cells in the system) and each cell computes its address coordinates in its own gradient-like proto-cell, as a function of its neighbours and its own state (**OK** or **not OK**).

2. *Division* - the step during which the mother cell begins to be divided into two descendents, in other words the genome transfer is started. The desired logical system is completely programmed on the cellular architecture after specialization and division

Some details with regard to cell differentiation are essential. The genes (instructions) are sequentially serially introduced one by one. Each cell loads the *f-gene* only if its *s-gene* corresponds to that *f-gene*. This second step process is also called *cell differentiation.*

The bus connecting part included in the function proto-cell is a network of **N**orth-**E**ast-**W**est-**S**outh connections: the cell receives the genome from its South and West neighbours, the neighbour being selected by a **JUMPER**, and sends its own genome to its North and East neighbours

3. But the embryological architecture may be stronger dedicated, possessing also the implemented *fault tolerance strategies*, (**BIST**) strategy most often. The architecture in this case it is more perforant, allowing *the third working step* if necessary.

- Each different functional part of the cell has implemented its own fault tolerance strategy, a **B**uilt – **I**n – **S**elf – **T**est (**BIST**) strategy or a dual rail of information, for example.

  By the (**BIST**) strategy in the RAM part of the nucleus-like proto-cell, for example, checksum bits are added to each memory word and a final gene codes the signature of the total memory.
- The immune system receives failure signals from different sub-blocks composing the cell or from the immune part of the neighbouring cells.

  As a consequence of the received failure signals, some cells become therefore *transparent* to the address computation; the first fault free cell located on its right takes its place (logical address).

Emergent Functionality Architectures. These architectures allow on-line evolution by real time development of a new functionality and new behavioural complexity for some autonomous agents typically in robotics or pattern recognition.

Two main types of such architectures are known: *co-operation architectures* and *subsumption architectures* (Shimonara 1994)

Co-operation architectures are characterized by the fact that many behavioural component subsystems are simultaneously active; there is neither subsumption, nor inhibition of any behavioural module by any other. The behavioural influences are considered together and the subsystem with most powerful influence is operational.

In case of ***subsumption architectures***, the behavioural modules inhibit one another as dictated by external environment. Subsumption is performed by behaviour pattern as mapping between the output of input sensors and input chromosomes for actuators (outputs of the system).

Evolvable Fault Tolerant Systems. This type of evolvable architecture is described in (Thompson 1995) and was evolved to control a real robot. It is an evolvable finite state machine held in *RAM* (see Fig. 2.12).

The signal representing the input and output variables runs asynchronously or is clocked. This choice is genetically determined by use of genetic latches controlling whether the signals are latched according to the clock or allowed to pass asynchronously. The clock rate is evolved too. Consequently the chromosome is a linear bit string that holds the *RAM* contents (the next state), the clock period and the status of each latch (present state, inputs).

This type of evolving system is a fault tolerant one: some type of faults affecting the phenotypes, a defective *RAM* bit for example, in the same way as would a genetic mutation, is better tolerated than in a non evolved system. In the above-mentioned evolving system the genetic encoding is made in a manner that the faults of interest have the same effect on the phenotype as the genetic mutation does it.

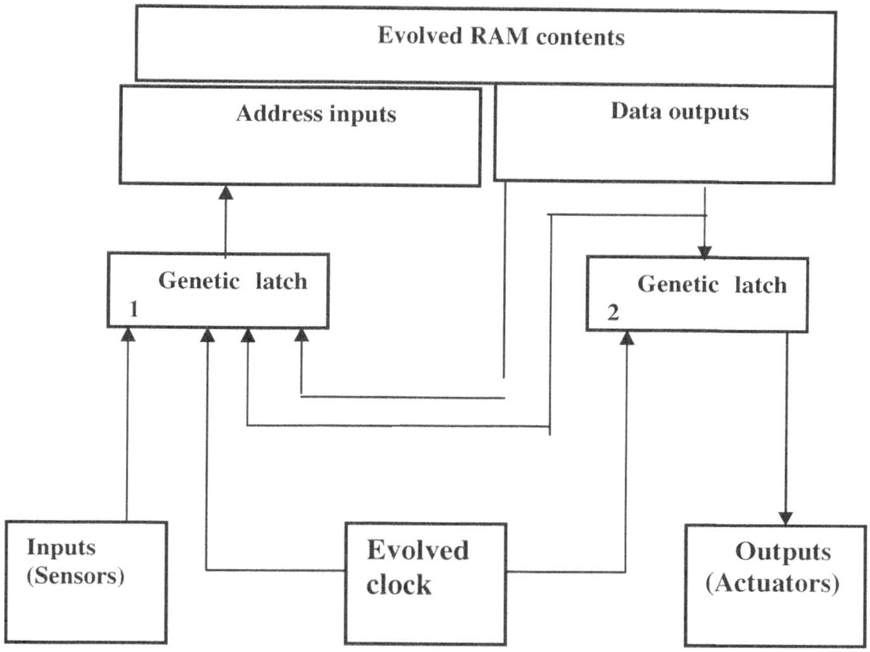

Fig. 2.12 Evolvable Dynamic State Machine Architecture after (Thompson 1995)

In case of a system working in an environment of fault, (Thompson 1995) suggests the use a co-evolving population of faults to help that the individual may be exposed to all faults of interest during their fitness evaluation, but this method was not sufficiently experimented. Also is prudently indicated to include the behaviour of the faulty part as a component of the fitness evaluation as a method to get fault tolerant systems in the situation of a defect persisting, but is said in the same time that the results in fault tolerance are not every time the expected ones.

Parallel Evolvable Architectures - of Higuchi Type. This type of architecture with intrinsic **EC** elements is a real time **RH** proposed by Higuchi (Higuchi et al 1994), (Higuchi et al 1996). We will call this architecture in this paragraph of the book as **EHW** .It has the greatest applicability nowadays: real time adaptivity of typical control systems in robotics or pattern recognition, for example.

EHW combines different *intelligent techniques* (**FL**, **NN**, **GA**) with **SC** *learning methods* (typically reinforcement learning or genetic learning), *on* a *parallel architecture.* This aggregation confers very interesting and unique properties to **EHW** architectures: real time adaptation as a reaction to the work of the external environment (by real time modifying the architecture's structure) and robustness (slow decrease of the performances due to the environmental perturbations or hardware faults).

The concept of **EHW** architectures includes three main component blocks built in the same massive parallel manner – each block composed by the same number of specific sub-blocks (see fig. 2.13):

- *the evolutionary component block* (**GA**) – a general structure, independent of application (a block of a parallel implemented **GA**, with **GA** operators acting bitwise, that means bit by bit action in parallel, on different chromosomes); by genetic operations, the **GA** block computes the architecture bits for **RLD** block.
- *the hardware reconfigurable component block* (**RLD**) - a general **FPGA-**relied structure, independent of application; this is a block of identical sub-blocks allowing real *time change of configuration and functionality* in hardware as a response related to the behaviour in the external environment; the outputs of **RLD** depend on inputs and the architectural bits from **GA**; the outputs of **RLD** are learned by the **RLC** component
- *the learning component block* (**RLC**); this block has an application dependent structure and usually is externally implemented; the **RLC** compute the *reinforcement signal* (reward or penalty) as an environment reaction to *modify the fitness function* of **GA** chromosomes

The typically performed function by **EHW** architecture is to learn the table representing the relation between the inputs and outputs, with application both in complex control systems of robots and in pattern recognition.

A hardware **GA**'s implementation can be tackled at different paralleled levels (Negoita 1995). But the compromise between the performances and the **GA** architectural complexity is the major task during this implementation - some limitations result naturally as an issue in hardware implementation as a consequence of decreasing the **GA** architectural complexity (Higuchi et al 1996):

-the suitable **GA** to be implemented is a canonical one
-the parent selection for the next generation is elitist
- genetic operators generating only one offspring must be implemented
any of the **GA** operation is applied
-the lifetime of a chromosome is limited to one generation (the entire population is replaced with the new generation).

There are three ways of **GA** *parallelization*:

1. Parallel on population and parallel on chromosome (all chromosomes in the population are simultaneously processed and genetic operators are applied for chromosomes on every alleles simultaneously)
2. Parallel on population and serially on chromosome (all chromosomes in the population are simultaneously processed and the alleles are processed serially)
3. Serially on population and parallel on chromosome (all chromosomes in the population are processed serially and genetic operators are applied for chromosomes on every alleles simultaneously)

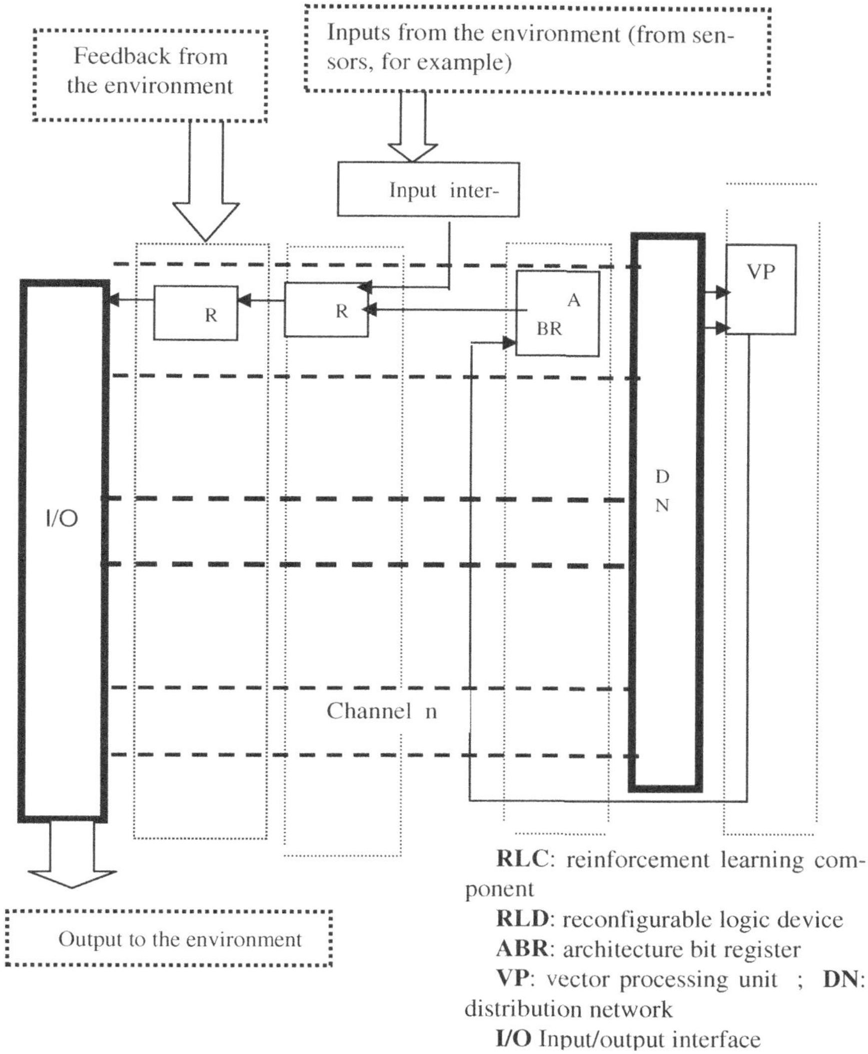

Fig. 2.13 The parallel evolvable architecture, after(Higuchi et al 1996)

Each of the above mentioned architectures may perform the proper processing speed but with different complexity costs.

The first type – is the fastest, but its complexity is the highest; the others two may perform smaller processing speeds, but with smaller complexity – these structures can realize the best compromise between performance and architectural complexity.

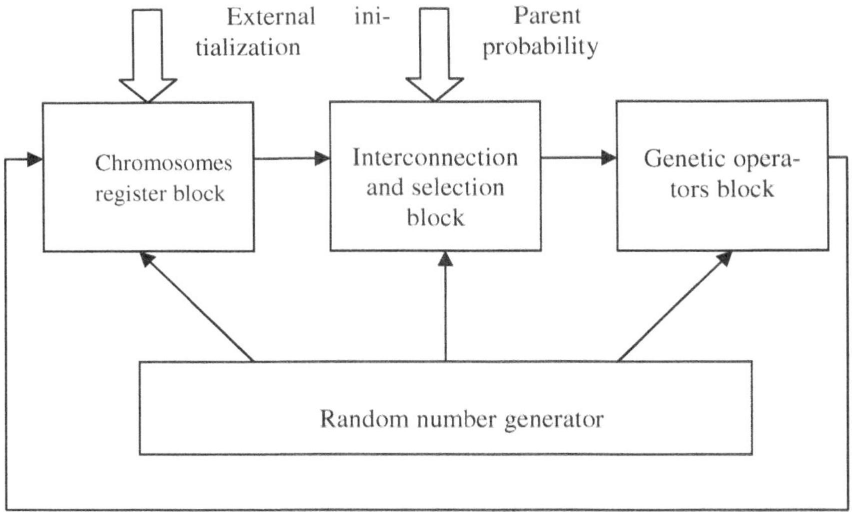

Fig. 2.14 The general configuration of a **GA** block for **EHW**

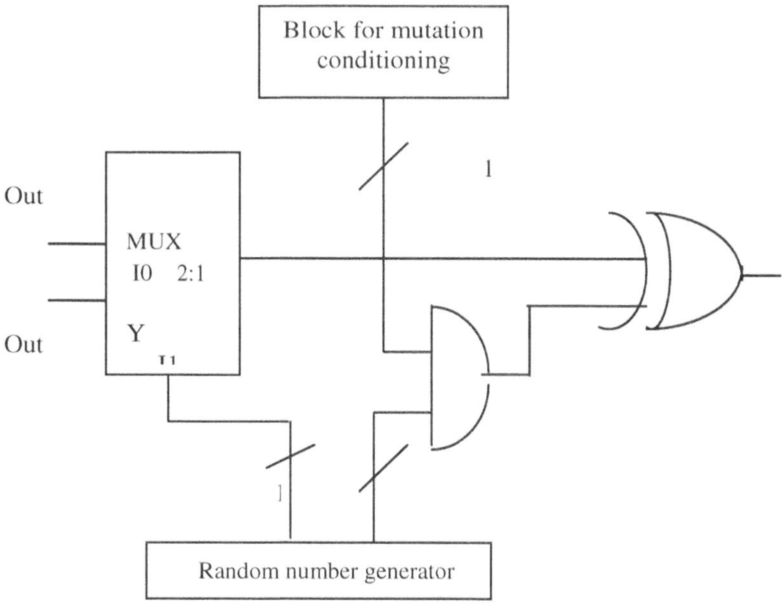

Fig. 2.15 Genetic Operators Block (channel *i*)

The **GA** hardware block of **EHW** has its general configuration (Mihaila Fagarasan Negoita 1996) as in fig.2.14, any the solutions for parallelization is adopted .

The processing architecture of the *serial on population and parallel on chromosome* type is adequate to a **GA** implementation on sequential computers. The processing *architecture with parallel processing of population and serial processing on chromosome* is a suitable solution to a **GA** implementation with long chromosomes. The main blocks have the following main building and functional features:

- the *block of chromosome's registers* may be *internal/external accessed* and *parallel/serially load*; its hardware configuration is built as an ensemble of *N* registers - *D* flip-flops shift-register namely (see fig. 2.16). Four hardware operations are performed as the case (*output operation* – the selection of the register bit corresponding to the allele that will be processed; *input operations* – selective writing of corresponding unique bit to the allele which was processed; *parallel loading* of all bits corresponding to a register - during the initialization stage; *serially loading* of all external bits corresponding to a register).
- *interconnecting and selecting parents block* contains *N* identical channels, each of them selecting one pair of bits belonging to the parents' chromosomes; its structure is a hardware multiplexing logic for the outputs $Out_{R,I}$ of the chromosome's registers
- *the genetic operators block* (see fig. 2.15 for one of its channels), is organized in form of *N* identical channels for one bit processing. Each channel performs a shuffle crossover (in the *MUX* section), followed by one mutation (in the *XOR* gate section). The sub-block for mutation conditioning has a similar functionality to the parents' selection sub-block.
- *the random number generation* sub-block both generates words of *L* bits length (where *L* - the alleles number in the chromosome) and generates the control words for other blocks.

See (Mihaila Fagarasan Negoita 1996) for more details with regard to main building and functional features of the blocks in the *parallel on population/serial on chromosome* **GA** processing architecture.

A modified **GA** called survival-based, steady state **GA** was developed in (Shacklefrod et al 2001) specifically for being suitable to an efficient hardware implementation both from the point of view of chip cost and **GA** operations speeds. A six-stage pipelined **GA** architecture was implemented for a survival-based **GA.** In *the survival-based steady state* a **GA** offspring survives only if it is more fit than the current least-fit parent – which it will then replace. This **GA** differs from the original **GA** where all offspring survive for a transfer to the new population.

Evolvable Architectures at Functional Level. All previously presented **EHW** architectures are using digital hardware, but their hardware synthesis was made at *a low-level hardware function* (of a *gate type* for example).

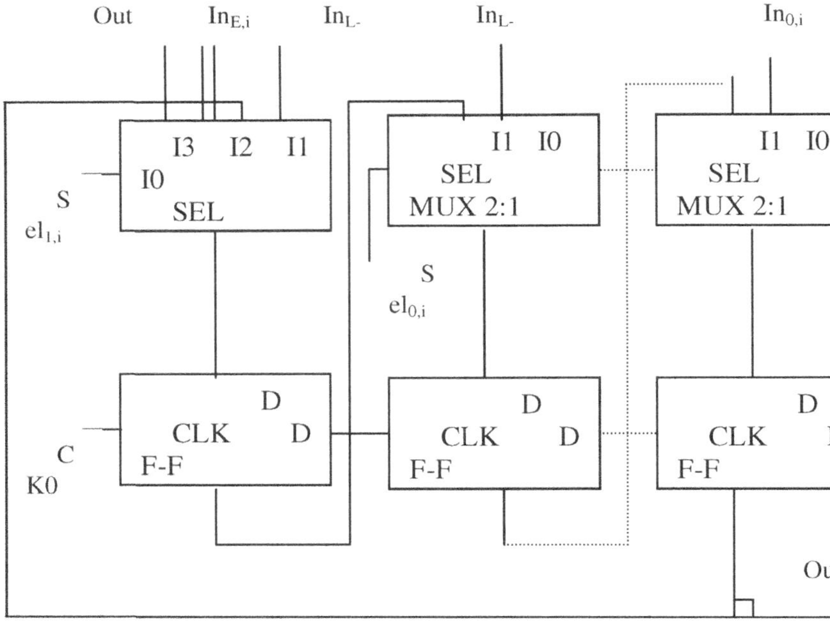

Fig. 2.16. The chromosome's registers block

The real necessity of evolving more complex applications can be solved by a genetically hardware synthesis at *a high-level hardware function* (of a *higher than gate type*).

This aim is feasible on a specialized dedicated **FPGA** architecture proposed by (Higuchi 1996) and called **F^{2}PGA**. This architecture is relied on a network of *switch settings* that allow the basic cells inside the device to be connected as the case. A basic cell in this kind of **FPGA** is called **P***rogrammable* **F***loating* **U***nit* *(***PFU***)* because of its availability of performing a large variety of functions: adding, subtracting, multiplication, cosine and sine using floating point numbers.

A **F^{2}PGA** programming word is a variable length chromosome containing both the **PFU**'s function programming and the crossbar switch settings. The chromosomes are stored in **A***rchitecture* **B***it* **R***egisters* (**ABR**) and the internal memory block stores the state of the cell system. This type of chromosome is used for high speed and flexibility in execution. The same speed motivation led to use of a **GA** handling the chromosomes by selection and mutation only. No crossover operator is used during the **GA**. The mutation operator may be of two categories: operand mutation (allowing for crossbar setting) or function mutation (allowing for **PFU** operation).

The following conclusion can be drawn on **EC** hardware implementation methods:

- all methods are based on biological metaphor inspired by evolution
- the applications are very different, from evolutionary methods of hardware design to real-time adaptation for complex control systems and pattern recognition
- advanced research is made in development in the world, partial results have been obtained, but the on-chip implementation is not yet achieved
- the most flexible and recently achieved applications are elements of behavioural intelligence of autonomous agents and back-up real-time adaptive hardware modules to the environmental situations including hardware malfunctions, the so-called fault tolerant systems.

The hardware implementation of **GA**, especially of those devoted to **EHW** architectures, arises some problems because of the required compromise between complexity of efficient **GA** and decreased architectural complexity that can solve the optimisation of hardware devices.

Good solutions of this problem seem to be tackling for the further research the topic regarding a simplified hardware implementation of efficient, but complex **GA** such as ***Nagoya*** **GA** (**NGA**) or even **GA** with variable length chromosome, the so-called **V***ariable* **L***ength* **G***enotype* **G***enetic* **A***lgorithm* (**VLGGA**). More details on **VLGGA** can be found in (Fagarasan and Negoita 1995), (Negoita Palade Neagu 2005).

Evolutionary design of digital circuits can be approached through different techniques as: gate-level evolution, circuit evolution in **PLA**, functional-level evolution, incremental evolution, evolution utilizing developmental schemes and even through some application-specific methods. None of these methods is a fully strength one fitting a general-purpose application target.

The practice of evolutionary circuit design revealed some typical practical issues: the bias in the applied design method; the relation of chromosome size versus complexity of circuits; the scalability of evaluation; the measurement of the level of innovation in an evolved circuit (Sekanina 2006).

Bias in the design method means the available initial set of gates that can be utilized and the available option in connectivity of these gates. In case of a circuit that is evolved at a functional level, the chromosome is shorter because the encoded circuit has fewer components and interconnections. This leads to a reduced search space – fast evolution, but a lack of innovation in the structure of the evolved solution. Long chromosomes, and a large search space, together with an increased level of innovation due to the chromosomes length, feature the gate-level evolution.

The relation of chromosome size versus complexity of circuits is known under the name of *scalability of representation*. It is the limitation in the complexity of the evolved circuits due to the available computing resources and the technological support. The size of the search space in a application of evolutionary circuit design is definitely influenced by the power of the computing machine.

The *scalability of evaluation* is a typical issue to the evolutionary design of the combinatorial circuits. In this particular case, the evaluation time of a potential solution (candidate circuit) grows very fast when the number of inputs increases largely. A strategy for counterbalancing the time of evalution is suggested in (Torresen 2002) , (Pecenka et al 2005).

The lev*el of innovation* in an evolved circuit is practically impossible to be measured, in fact it is regardless the selected approach of evolutionary circuit design and also does not depend on the complexity of the application (the evolved circuit).

AEH technology performed relevant achievements in providing self-reconfigurability, adaptability and evolvability to different applications. But the self-reconfigurable architectures are still featured by not enough real-world efficiency regarding several main parameters the evolved circuit design offers to them. These limitations are either in technological parameters of the evolved circuit - the reconfiguration times, the power consumption, or because of the technical design support – computer resources and the intrinsic properties of the silicon environment, or electronics manufacturing indexes of performance – cost, predictability and safety/reliability.

An honest comparison of **AEH** technology with its main rival **ASIC** technology is useful to the **EHW** practitioners. At a first glance, **ASIC** must be preferred because its philosophy consists of a distinct hardware implementation for each required function. **AEH** combination of **RH** and **RM** leads to a larger overhead. This preference for **ASIC** is counter attacked by the fact that **ASIC** proves to be still an inefficient economical solution (Stoica and Andrei 2007).

RM – the software component of the **AEH** – employs **EA**, but the huge dimension of the search space and the intrinsic required fitness evaluation make the search difficult indeed. The total **EHW** design time for a circuit is high because of summing the high number of evaluation (number of iterations times number of points in one iteration) and the relatively long time spent for fitness evaluation at each iteration. This inefficient transition mechanism reflects by excessive consumption of time, power and computational resources. The reconfiguration time is greatly improved in the latest Vitex devices providing enough fast clocks and supporting partial reconfiguration for a large range of applications. The requirements of other applications, as the adaptive computing are still not covered. Ways of possible solutions of this problems are presented in (Torresen 1999) and (Zebulum et al 2007).

AHS act in the real world and are faced to cross through unpredictable, undesirable intermediate states that are unsafe and real dangerous for system survivability. If undesirable states are encountered, **AHS** integrity is drastically endangered in situation when hysteresis is used in loop control of a system. Possible solutions are suggested by (Stoica Andrei 2007).

Another factor that diminishes the efficiency of self-reconfigurable **AHS** architectures is the lack of no dedicated tools and procedures for verification and validation. Currently , the solution is guaranteed only when tested, but an accurate confidence in the evolved circuit solution must be ideally proved in real time during the adaptation / self-reconfiguration process.

2.1.4 An Application of GA for the Design of EHW Architectures

The real-time adaptation in **EHW** systems means changes in their architectures, as a reaction to the variable external environmental conditions and as a privilege conferred by the properties of the technological support of these circuits. This approach is a step further than the standard approaches so far, where adaptation only affects software structures. The application discussed in this part of the book treats a **GA** with local improvement of the chromosomes, used to optimise the evolved configuration of **EHW** architecture of Higuchi type got as a result of a **GA**. It is presented the simulation of the **EHW** circuits behaviour in the framework of the objective function used in **GA**. This kind of application appears in robotics, namely for characterizing the behaviour output pattern specific to an autonomous agent as a mapping problem between sensors outputs and actuator commands.

Comparative experimental results in case of an *XOR* objective function with *4* inputs are presented using both a classical Goldberg **GA** and the above-mentioned type of **GA**.

RLD is a principal component part of **EHW** Higuchi type architectures. As previously mentioned, the general structure of **RLD** block is based on the technological support of **FPGA**. One of the most used **FPGA** is **PLD** - a structure that is composed of a lot of identical macro-cells.

A *GAL16V8* **PLD** circuit (like those delivered by Lattice Semiconductors) was used to develop a simulation for an application to be introduced in this book chapter. *GAL16V8* is a typical **PLD** circuit composed of *8* macro-cells, each of them having *16* inputs and 8 outputs. The simplified structure of a macro-cell in *GAL16V8* is presented in Fig. 2.17.

This macro-cell has *16* inputs for the input variables delivering the *AND* terms of the *8* columns (each of the input variables may be or not be negated on the inputs of the logic gates). The *8* columns deliver the *OR* terms of the logic function at the output of the macro-cell. The block of architecture bites – the block of binary command words for both architecture and macro-cell function changing – was omitted for the simplicity of the figure.

The **RM** tool selected for this application was a **GA** *with local improvement of chromosomes.* The main reason of local improvements in chromosomes is just the application of *genetic operators at the level of each group gene* in the chromosome.

GA *with local improvement of chromosomes* has a step sequence as described in (Furuhashi et al 1994) and also (Yang et al 1995):

Step 1 – random generation of an initial population (each of the individuals is randomly generated as in a classical **GA**);

Step 2 – applying a specific operator, called *NAGOYA mutation*, during a generation, generates the new population of chromosomes

2a) *chromosome implementation as a group of segments*

- each of the chromosomes has *N segments (group genes)* – each "specialized gene" may be separately evaluated in the external environment. Each of the *group genes* (specialized genes) in a chromosome is selected one by one and reproduced into a pre-established number of copies;

- o the mutation operator is applied to each set of these *group genes*; each of this mutated set of specialized genes is evaluated in the external environment, of course in the frame of the whole chromosome; the best-evaluated group gene is selected as a chromosome segment. All the *N* segments of the chromosome are implanted in this manner.

2b) this genetic operation called *NAGOYA mutation* is applied for getting all chromosomes in the population, one by one.

Step 3 – the population representing the *new generation* is got as a result of a typical operation chain performed by an usual **GA** to the population got at *step 2* ***after Nagoya mutation was applied***. This operation chain consists of the parents selection, followed by the reproduction, crossover and the replacement operation, this last operation meaning the selection of the new individuals from the parents and the children.

The algorithm continues after the end of the *Step 3 for* a new generation from *Step 2*

The object of application in this sub-chapter is targeted to obtain a connection matrix of the **PLD** circuit as suitable as possible for matching the objective function represented by the inputs/outputs mapping the system including the **EHW** architecture. This problem was solved at the level of one macro-cell by a program written in high level language (Negoita and Dediu 1997) and (Negoita 1997). The solution at the level of the whole circuit as the case is an accessible extension.

Some deeply details on practical problems that might be met during selection of a suitable **GA** encoding for a particular application will be introduced at this point too. *A* chromosome *encoding the problem* solution required by this application case has *8x16x2* binary values represented by concatenation of the possible connection points codifications for all the *8* columns in the connection matrix. The above-mentioned encoding uses four states for a connection point in the matrix: *A*; *not A*; *0* – for simultaneously connected inputs *A* and *not A*; *1* – for disconnected inputs. It is a convenient but redundant coding leading to a search space of 2^{256} search points. In this form of codification, the points (potential solutions) in the search space got in case of a connection point with a logic state *0* are specific to the *AND* gates outputs, no more adequate to be considered as *OR* inputs. These illustrative points for simultaneous *A* and *not A* inputs connection led as a consequence to a tri – state coding in a search space of $3^{16x8}\ x\ 2^{8}$ potential solutions.

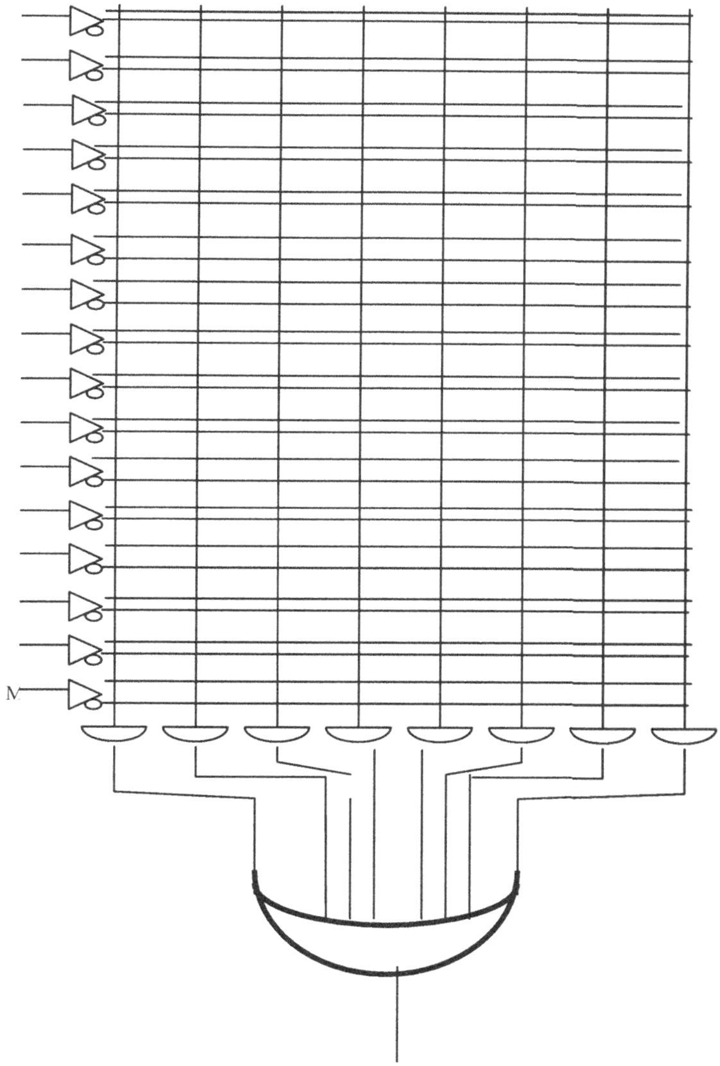

Fig. 2.17. The simplified **PLD** macro-cell of GAL16V8

An economy *Ec* is got by a tri – state coding of the search space as follows:

$Ec = 2^{256} : (3^{128} \, x \, 2^{8})$ $\qquad$ $Ec = 2^{248} : 3^{128}$

$2^{10} = 1024 > 10^{3}$ $\qquad$ $3^{2} = 9 < 10$

$2^{248} > 10^{24.8x3}$ $\qquad$ $10^{64} > 3^{128}$

$Ec \cong 10^{74.4-64} \cong 10^{11}$ search points.

A tri – state codification to obtain an economy of 10^{11} points in the search space of the solutions is more suitable for searching the solution space.

Some deeply details on practical problems that might be met during design of a relevant fitness function for a particular **GA** application will be introduced at this point too.

The concrete application was made on a mapping problem represented by an *XOR* objective function with *4* inputs x_1, x_2, x_3, x_4. The disjunctive canonical form of this function is as follows:

$f =$ *not* x_1.*not* x_2.*not* x_3.x_4 **OR** *not* x_1.*not* x_2.x_3.*not* x_4 **OR** *not* x_1.x_2.*not* x_3.*not* x_4 **OR** *not* x_1.x_2.x_3.x_4 **OR** x_1.*not* x_2.. *not* x_3.*not* x_4 **OR** x_1.*not* x_2.x_3.x_4 **OR** x_1.x_2.*not* x_3.x_4 **OR** x_1.x_2.x_3.*not* x_4

This logic function *f* has *8* terms, that means it is suitable to be implemented on one *GAL16V8* macro-cell.

Let us note: *I*- the input logic variable; *c* – the logic variable representing the command (input) bit of the equivalent switch to a connection point in *GAL16V8* cellular connection matrix (*0* – open contact; *1* – closed contact). Then the truth table of the logic function *f* in the connection point of cellular connection matrix is as in Table 2.1

Table 2.1 The truth table of the logic function *f*

I	c	f	not f
0	0	1	0
0	1	0	1
1	0	1	0
1	1	1	0

Accordingly, the above-described logic function is $f =$ *not c* **OR** *I*. This relation is nevertheless available for the combinations of parallel type as *matr[I].v*, $input_{16}$, and *matr[I].vnot, not* $input_{16}$ too.

The operation *not c* **OR** *I* is applied as a consequence, in parallel (*16* bit) to calculate the f_c function for these above-mentioned parallel combinations.

In case of one of the *8* **AND** columns ($I = 1 \div 8$) has the output *1*, *AND* := *1*, this value *1* is automatically assigned to the **OR** output, *OR* := *1*. Otherwise all the **AND** columns ($I = 1 \div 8$) are calculated and the **OR** output is assigned as *OR* := *0*. After calculation of *OR* := *0* or *OR* := *1*, the relation between f_c and f_{req} it is checked up:

$f_c = f_{req}$ leads to $err(input_{16}) = err(input_{16})$ – unchanged error;
$f_c \neq f_{req}$ leads to $err(input_{16}) = err(input_{16}) + 1$, the incremented error,

where f_c, f_{req} are the calculated.

The requested value of the objective function f for our **GA** application

$err(input_{16})$ is the difference between f_c , f_{req} .

The evaluation (fitness) function of the chromosomes used in this **GA** represents in fact a simulation of the **EHW** circuit requested to get the objective function on a *GAL16V8* macro-cell. It is a scalar that represents the total number of errors between fc , f_{req} values of f, for all possible combinations of input variables:

$$f = \sum_{input16=0}^{\max input} err(input16)$$, where $input_{16} = 0 \div maxinput$ are the values corresponding to all the possible parallel combinations of the input variables.

f_c is calculated for a combination of input variables with the following logic assignment formula:

$$\mathbf{fc} = \mathop{\mathbf{OR}}_{i=1}^{8} [(input_{16} \ \mathbf{OR}\ not\ \ matr[I].v)\mathbf{AND}(not\ \ input_{16}\ \mathbf{OR}\ not\ \ matr[I].vnot)],$$

where:

$input_{16}$ – the parallel inputs combination; *not $input_{16}$* – the parallel combination of complemented (inverted) inputs

matr[I].v – the parallel combination of command bits for the connection points (on the matrix column) corresponding to the non negate variables;

matr[I].vnot – the parallel combination of the command bits for the connection points (on the matrix column) corresponding to the negated input variables.

Experimental results were presented in (Negoita and Dediu 1997) and (Negoita 1997). A detailed comparison of the solutions found to this application problem by using both a classical Goldberg's **GA** (**SGA**) and a **GA** of Nagoya type (**NGA**) may be found in the two works.

SGA has used a roulette- wheel selection and the new generation was got by a total replacement of the old generation.

NGA used in the works has $m = 5$ – the *copies* number for a group gene, and the number of *group genes* (*segments*) is $N = 8$, namely the same as the number of columns in the connection matrix. In contrast with the *Nagoya algorithm* used in (Furuhashi et al 1994) and also in (Yang et al 1995): after applying the *Nagoya mutation* on all chromosomes, *Step 3* of our (improved) **NGA** includes the mutation too, not the reproduction and crossover only, the parents selection type and the replacement of the population for the new generation being kept.

Both **SGA** and **NGA** were ran *10* times (runnings) for *4* common settings of the **GA** parameters:

A) $p_c = 0.8;\ p_m = 0.05$ *population = 40; no. of generations = 400*

B) $p_c = 0.4;\ p_m = 0.001$ *population = 40; no. of generations = 400*

C) $p_c = 0.4;\ p_m = 0.001$ *population = 40; no. of generations = 1000*

D) $p_c = 0.8\ ;p_m = 0.001$ *population = 40; no. of generations = 1000*

Table 2.2 Experimental comparison of the **SGA** and the modified **NGA** for the application case

Err	Rng. SGA	Rng. NGA	Rng. SGA	Rng. NGA	Rng. SGA	Rng. NGA	Rng. SGA	Rng NGA
6			4				2	1
5		1	4	4	8	2	5	1
4	3	6	1	5	1	4	3	4
3	6	3		1	1	4		3
2	1							1
Ev	16 x 10^3	64 x 10^4	16 x 10^3	64 x 10^4	4 x 10^4	16 x 10^5	4 x 10^4	16 x 10^5
pc	0.8	0.8	0.4	0.4	0.4	0.4	0.8	0.8
pm	0.05	0.05	10^{-3}	10^{-3}	10^{-3}	10^{-3}	10^{-3}	10^{-3}

The meaning of notations in Table 2.2 are as follows:

- *Err* – the error values;
- *Rng* – the running number;
- *Ev* – the evaluations number;
- p_c – the crossover probability
  p_m – the mutation probability

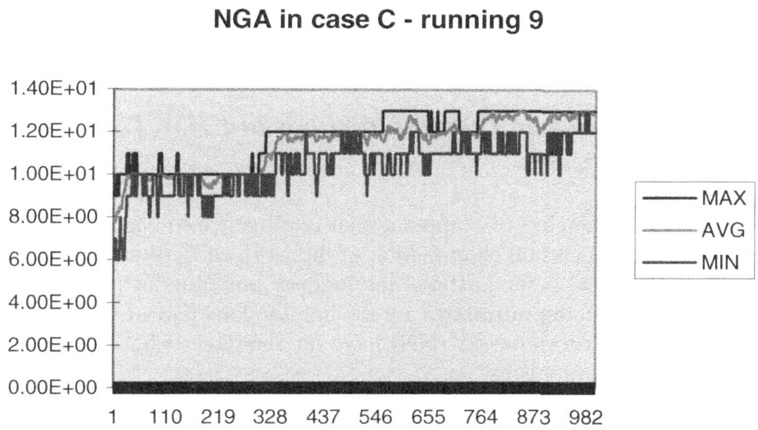

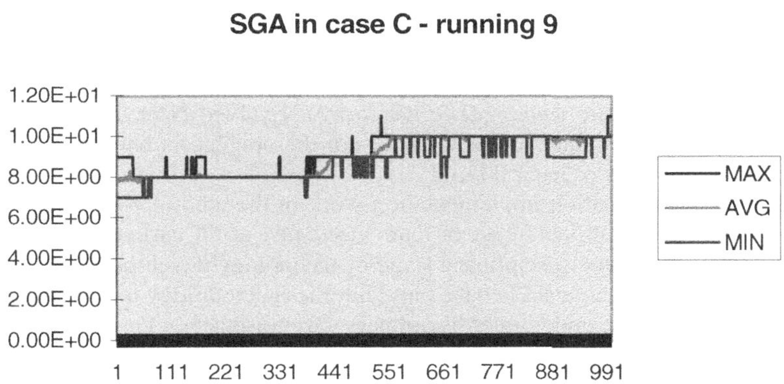

Fig. 2.18 A graphical example for both SGA and NGA

NGA leads to better results in *3* of the *4* cases in this table of results. The comparison criterion was by getting the minimal error in as much as possible running and using one of the least suitable selection for the new generation, namely the total replacement.

Figure 2.18 illustrates running *9* in case *C* both for **SGA** and for **NGA**, namely representing the *maximum value (MAX), the minimum value (MIN)* and the *average value (AVG)* of the objective function error in the population (vertical – axis) depending on the generation number (horizontal – axis).

The experimental results illustrate the **NGA** superiority given the **SGA** for application in design elements of **EHW**, despite of the fact that the number of fitness function evaluations was slightly bigger. This superiority was proved despite of keeping the total replacement for the new generation in **NGA**. But better results will be obtained certainly by using another type of selection for the replacement, an elitist one for example, allowing the best individuals from the previous generations to be kept.

2.1.5 Global Remarks on Current Methods in EHW Technology and Its Prospectus

The analogy of some branches of computational intelligence to Natural Life must be the starting point of a careful examination of the work carried out to date under the heading of **EHW**: **EC** is the artificial intelligence homologous part of natural *phylogeny*, **self-reproducing automata** are the homologous part of natural *ontogeny* and neural and *neural networks* (**NN**) have an analogue behaviour to epigenetic natural process in organisms.

The bio-inspired systems can be built and classified inspired by the powerful natural examples as following the above-mentioned levels of organization in Natural Life. The same considerations are applicable to **EHW** that may be viewed by its definitions as a sub- domain of artificial evolution.

The *phylogenetic hardware* is of different types, but any this type is, it deals with evolutionary algorithms applied to the synthesis of digital and analogue electronic circuit. Different self-reproducing hardware methods are illustrating the self-reproducing hardware (*ontogenetic hardware*). Evolved **NN**s, including **NN** structures based on **C**ellular **A**utomata (**CA**) are the epigenesis hardware implementing the architecture of natural brain.

Our R&D and application implementation work in the area of **EHW** must run under idea that a practitioner must be convinced just at an earlier professional stage about the actual interdisciplinary stage of technology development. So they must get not the knowledge/skills base only, but the methodology too of acting in real world applications under interdisciplinary circumstances. This is available nowadays not for recent graduates only, but for other practitioners also (engineers, scientific researchers and academic staff) any their background and area of interest is. The progress of **EC** applications in the framework of **SC** toward **CI** must be discussed from an application - engineering point of view rather than from a cognitive one.

The most spectacular **EC** application in the **SC** framework is the **EHW** in the frame of **AHS** nevertheless. **EHW** opened a revolutionary eve in technology and in social life development by its radical implications on engineering design and automation (Negoita Neagu Palade 2005):

- a dream of humanity became reality - the *systems adaptivity* was implemented (transferred) by **EHW** *from software to hardware*
- a drastic time saving way from design to real world application of intelligent hardware is used, *no more difference* exists *between design* and *adaptation* concerning **EHW** based machines having a *behavioural computational intelligence*

- *electronics engineering was fundamentally changed* as a profession by using **EHW** custom design technologies instead of soldering based manufacturing.

Currently, many complex real-word processing applications are implemented on general-purpose processors as Pentium. However, the complex and fast computation has recently started to be performed by dedicated hardware instead of software in digital computers because hardware can operate in parallel successfully. The **EHW** technology as the co-work of **RH** and **RM** is the best way of implementing **AHS** able to reconfigure itself to achieve a better performance.

An **EHW** system of high performance in very instable environments is presented in (Jeon et a 2005) for an adaptive image processing of images in space. The **RH** module is implemented on advanced **FPGA**, namely using Xilinx Virtex 2 and processes the median, histogram equalization, contrast stretching and illumination compensation algorithm. The **RM** evolvable software is a hybrid **GA** and feature pace search, consisting of a **G***enetic* **A***lgorithm* **P***rocessor* (**GAP**) and **ARM** core - the first-ever processor core designed for synthesis onto the programmable elements of **FPGA**s. **ARM** architecture (previously, the **Advanced RISC Machine**, and prior to that **Acorn RISC Machine**) is a 32-bit **RISC** processor architecture developed by ARM Limited. Help of a fitness function performs the adaptation of this **EHW** system. The fitness function is defined for the system performance and the class scattering criterion. The system performance involves the correctness that the evolvable adaptation has achieved so far, and the class scattering reflects the expected fitness on future generations. The system adaptivity is proved by the fact that proposed **EHW** performs very well regarding external changing illumination and various noises.

After a deep examination of the framework involving *bio-inspired (hardware) systems*, another important concluding remark arises: a strict adherence to solutions in nature is not the optimal solution to these systems, deviations from what is strictly natural are no doubt of use in hardware bio-inspired systems. Even other environments are seen as possible evolvable media in nearest future: *wetware* – real chemical compounds are to be used as building blocks or *nanotechnology* – relied on molecular scale engineering. Much more, the hardware bio-inspired systems hosted by such environments might embed evolutionary, reproductive, regenerative and learning capabilities together.

Different practitioners studied the possibility of using other medium for the **EHW** (Miler and Downing 2002), (Harding and Miller 2004).

(Miler and Downing 2002), try to justify the strategy of using artificial evolution for direct capitalizing the properties of materials at a molecular level, namely through materials having associated no electronic functions and allowing artificial evolution to work outside the constraints featuring silicon based technology. They think to possible advantages that extremely promising non-conventional technologies (no silicon medium based environments) may confer to **EHW** over the conventional electronic technology (silicon based medium). They sketched an evolutionary exploitable device called the **C***onfigurable* **A***nalog* **P***rocessor* (**CAP**) and also thought how it might be individually trained.

One possible **CAP** configuration was as a **F***ield* **P***rogrammable* **M***atter* **A***rray** (**FPMA***), an architecture inside where applied voltages may induce physical changes that interact in unexpected ways with other distant voltage –induced configurations in a rich no silicon physical substrate. A second variant of possible **CAP** configuration was as an amorphous cellular automatonhaving its local rules determined by the local configuration and whose cells may be connectable to a number of distant cells or may be non-linear coupled.

The **CAP** suggested material core might be: the **l***iquid* **c***rystal* (**LC**), conducting and electroactive polymers or voltage-controlled colloids – suspensions of particles of sub-micron sizes in a liquid. (Miler and Downing 2002) concluded that **EHW** type intrinsic evolution may be even best attempted in physical substrates that are rich and complex of the above mentioned materials than in conventional silicon transistor based substrates.

The possibility of using switchable glass technology -also called *smart windows* – for **EHW** – like adaptivity is suggested in (Oltean 2006). Switchable glass is a technology reliant on using electrical power for controlling the transmission of light through windows .The electrical power is used through applying variable voltages to the window and the effect is that the amount of transmitted light can continuously vary. This kind of technology is suitable just for light-based computation only because switchable glass involves and affects only the transmitted light.

The characteristics that **EHW** candidate materials should have are suggested in (Harding and Miller 2004). A possibility of potentially using switchable glass as a platform for **EHW** tasks was suggested because of such kind of material would have all the required properties to **EHW** candidate materials. These characteristics are in case of switchable glass technology as follows:

- this material can change its degree of opaqueness by applying some electrical power
- this material affects the quantity of light that pass through it
- this material is able to be reset to its original state by removing the source delivering the electrical power

The implementation of the proposed hardware was not yet achieved, but the switchable glass technologies were used in an experiment where an **EHW** device was used for computing the fitness of a chromosome. The used **EHW** device consisted of the following parts: several continuous sources of light; an array of **LCD** cells; an array of **SPD** cells; some biconvex lens to focus all rays in a given point; a photo cell which captures the light to be converted into electric power; finally this electric power is sent to a computer in form of an analog signals that is used for fitness processing.

A decent global view is useful to **EHW** practitioners on the whole range of approaches proposed to the evolutionary circuit design and on the levels of complexity and innovation that these approaches may confer to the required application.

The real-world applications field does not require an absurd accurate handling of **EHW** classification, terminology and definition. But the final front-users efficiency is reliant on making the distinction between two **AHW** approaches: evolutionary circuit design and **EHW.** This is absolutely justified by their radically different technical objectives.

Evolutionary circuit design is an unconventional method aimed to implement better intrinsic parameters (area, speed, power consumption, thermal dissipation) than the conventional design methods and/or even to confer original features to the circuitry - self-organization, fault-tolerance, **BIST** testability.

EHW is hardware reliant on cooperative work of **RH** and **RM** for implementing high-performance adaptive applications starting from very sharp (incomplete) problem specifications and environment-aware (able of survival and recover from unpredictable and dangerous events).

An honest comparison of **AEH** technology with its main rival **ASIC** technology is useful to the **EHW** practitioners. At a first glance, **ASIC** must be preferred because its philosophy consists of a distinct hardware implementation for each required function. **AEH** combination of **RH** and **RM** leads to a larger overhead. This preference for **ASIC** is counter attacked by the fact that **ASIC** proves to be still an inefficient economical solution (Stoica and Andrei 2007). Currently, **AEH** is the most preferred solutions for academic research or for very specific applications, but the demand for general purpose and customer-designed circuits is still the monopole of **ASIC** technology. The maximal impact in advancing **AEH** is still not achieved by reducing cost per function and reducing overhead. But **AEH** technology is strongly relied on hardware flexibility at various level of **RH**, its reprogramming speed and algorithmic efficiency is in improving, the embedded intelligence – self-reconfiguration - is relevant to the finest level of the hardware. These are serious reasons for a compromise between the economic aspects and the technological efficiency as a general rule for most hardware implementation. But the work under this compromise must be certainly abandoned, if the stringent application requirements ask for self-reconfigurability, adaptability and evolvability imposing the use of **EHW** in favour of **ASIC** in some applications regardless the business overhead.

2.2 Hardware Implementation of the Artificial Immune Systems

This section is not supposed to be an absolutely complete review of **AIS** hardware implementation field with detailed descriptions, but mainly focus on some aspects of **AIS** hardware implementation in context with the **HIS** framework that progressed toward some spectacular and effective combinations envisaging even the emergent **CI** paradigms such as **DNA** *Computing* and **EHW** (Negoita Neagu Palade 2005). The artificial immunity-based systems are hybrid information systems almost by their nature as from the three main informational features (Ishida 1996) ,(Ishida 2004), (Dasgupta and Attoh-Okine 1996), so they offer an optimal technical frame of hybridization with all other **CI** paradigms both at the level of model and most important at an algorithmic level (Negoita 2005) and (Negoita Neagu Palade 2005). (Bradley and Tyrell 2001).

A recent intelligent hybridisation of **AIS** is applied in case of one of the most revolutionary technology nowadays, namely in case of **EHW**. A main reason for **EHW** – **AIS** hybridisation was reliant on two **AIS** features, healing and learning, that were applied to design **EHW** fault – tolerant **FPGA** systems (Bradley et al 2000). An additional layer that imitates the action of antibody cells was incorporated to the previously elaborated embryonic architecture by the same team (Ortega et al 2000). Two variants of this new **EHW** architecture use an interactive network of antibody cells featured by 3 independent types of communication channels: the data channels of the embryonic array of cells, the data channels of antibody array of cells and the inter-layer communication channels ensuring that antibody cells can monitor the embryonic cells. The antibody array of cells performs monitoring and checking of the embryonic array of cells, so that the correct functionality of any particular **EHW** configuration can be achieved at any time.

Another **AIS** inspired variant of **EHW** hardware fault detection was reported in (Bradley and Tyrell 2001). They used an advanced **FPGA** hardware- Virtex XCV300 - to implement a hardware negative clonal selection **AIS** attached to a **F***inite* **S***tate* **M***achine* **(FSM).** This is very important because any hardware system can be represented by either a stand-alone or an interconnected array of **FSM**s. This work presents how a **FSM** can be provided with a hardware immune system to perform a novel form of fault detection, giving the ability to detect every faulty state during a normal operating cycle, The authors called this novel fault detection as *immunotronics*.

A software **AIS** has been implemented for robot error detection to identify all faults that generates an error greater than a pre-specified limit (Canham et al 2003). The **AIS** learns normal behaviour during a fault free training period. Following the learning period, the error detection was made for two different kinds of robots, namely either for a controller and for the motion control manager. The **AIS** was an excellent error detector proving to be independent to the system under test. Although being implemented in software, this **AIS** has been designed to enable a hardware implementation within a **FPGA** or embryonic array.

A **DNA-AIS** intelligent hybrid system was reported in (Deaton et al 1997) where **DNA** *Computing* was proved as an alternative to hardware implementation of **AIS**. In this work, an **AIS** negative selection algorithm was implemented in a **DNA** computing framework, using *in vitro* techniques from molecular biology. Using **DNA** single strands under denaturation, renaturation and splicing operators, the censoring and monitoring parts of this selection algorithm were successfully implemented.

But **AIS** is not the unique **CI** technique that was implemented by biomolecular **DNA** through *in vitro* methods.

Also some elements of **FS**, in particular the fuzzy membership functions and a fuzzy associative memory (Deaton and Garzon 2001) have been implemented by mechanisms of molecular computing.

The molecular mechanism for the implementation is **DNA** hybridisation , in which two singe stranded **DNA** molecules – oligonucleotides – bind to form a double-strand duplex. **DNA** hybridisation is intrinsic fuzzy and as a consequence, there is a degree of uncertainty as to weather the **DNA** molecule is in the double-stranded or single-stranded state. The uncertainty in this hybridisation is handled hence to implement fuzzy variables and fuzzy inferences.

2.3 Hardware Implementation of DNA Computing

The original interest around the idea of **DNA** computing was caused by its potential to solve computational problems through massive parallelism. However, the current technological support is far from being capable of the level control of molecules that is really required for large, complex computations.

The fully hardware implementation of **DNA** is a very difficult target. But **AHS** techniques are used for hardware implementation of some **DNA** Computing elements.

The biological sequence alignment task in the field of computational biology uses the **S**_mith_-**W**_aterman_ (**S-W**) algorithm as a biological alignment sequence. A scalable accelerator for both **DNA** and protein pair-wise sequence alignment is presented in (Abouellail et al 2007). This algorithm was implemented by multi-node **H**_igh_ **P**_erformanc_e **R**_econfigurable_ **C**_omputers_ (**HPRC**s) for computational biology applications, namely two reconfigurable platforms – _SRC-6_ and _Cray-XD1_ were used.

Two reasons justify the suitability of this algorithm for acceleration with **HPRC**: first, it can be massively paralellised (both fine-grain and coarse-grain parallelism was exploited); second, it is a good candidate to address a wide variety of **HPRC** architecture issues, including the scalability when the number of nodes increased. The size of data for this application is quite large, but the available local memory for **FPGA** is still small. This is an issue of the method causing very large databases and queries request for sufficient larger **FPGA** to solve them.

An example of the support that **CI** techniques are able of providing to **DNA** Computing was presented above.

On its turn, the molecular approach may help new implementations of the **CI** techniques.

A new implementation of **G**_enetic_ **P**_rogramming_ (**GP**) based on dataflow techniques in **DNA** Computing is presented in (Wasiewicz and Mulawka 2001). Structures of the directed graph describe a program in the data flow unit of the data flow computers having fully parallel architectures. In the dataflow computers **GP** may be executed on the populations of graph-structures describing logical functions. The authors of the above-mentioned paper consider representation of the directed graph by means of **DNA** molecules. They handle these graph encoding molecules in a way, which can be considered a **GP** algorithm. Logical function arguments are treated as data that are input signals of the dataflow function graph consisted of special and arc strings. The data flow machine structure is dependent on results of **GP** operations and changes after each cycle. During execution of the **GP** algorithm the operation of amplification is performed only in one **DNA** test tube and the identification of its products is performed by electrophoresis.

The approach provided useful in development of new powerful parallel computers (Gehani and Reif 1998).

The possibility of **NN** to help **DNA** Computing was presented in (Mills et al 2001) .They considered that a **NN** in which the usual axons and neurons are replaced by the diffusion and molecular recognition and explored this as a possibility toward fault-tolerant molecular computation. The experimental aspects helped for estimations regarding the ultimate speed of **DNA** Computations.

Population-oriented **P***article* s*warm* **o***ptimisation* (**PSO**) is a bio-inspired technique that has been proposed for supporting the **AS** application of dynamic adaptation of array antennas (Kokai et al 2006). But no **PSO**-based methods have been used to perform complete evolution of digital hardware systems, despite the **PSO** suitability to population-oriented implementations, since their population size are usually small.

(Pena and Upegui 2007) achieved a population-oriented hardware implementation of the **PSO** with **PSO-DR**. This hardware implementation is featured by a **RAM**-based genotype-phenotype mapping, distributed and parallel processing capabilities, distributed storage, pipelined data paths and fixed point arithmetic. The design is synthetizable on advanced **FPGA**, namely on Xilinx Virtex –4.

2.4 Elements of Intercommunications Inside the AHS/EHW International Community (Conferences; Books; Journals; Elite Departments)

AHS/EHW Conferences

NASA/ESA Conferences on Adaptive Hardware and Systems
Most recent events:

2007 NASA/ESA Conference on Adaptive Hardware and Systems, 5-8 August 2007, Edinburgh, UK
General Chair: Tughrul Arslan, University of Edinburgh, UK
2006 NASA/ESA Conference on Adaptive Hardware and Systems, 15-18 July 2006, Istanbul, Turkey
General Chair: Adrian Stoica , NASA Jet Propulsion Laboratory, USA
2005 NASA/DOD Conference on Evolvable Hardware, July 2005, Washington DC, USA
General Chair: Jason Lohn, NASA Ames Research Center
ICES International Conferences on Evolvable Systems: From Biology to Hardware

Most recent events:

The 7th International Conference on Evolvable Systems: From Biology To Hardware (ICES'07), September 21-23 2007, Wuhan, China
General Chair : Lishan Kang, China University of Geosciences, China

The 6th International Conference on Evolvable Systems: From Biology to Hardware (ICES 2005) Barcelona, Spain, September 12-14 2005
General Chair : Juan Manuel Moreno Arostegui, Universitat Politecnica de Catalunya, Spain

Conferences having AHS/EHW as a main stream

KES International Conferences on Knowledge-Based & Intelligent Information & Engineering Systems
Most recent events:

KES2007 11th International Conference on Knowledge-Based and Intelligent Information & Engineering Systems, September 12-14 2007,Vietri sul Mare, Italy
General Chair: Bruno Apolloni, University of Milano, Italy
Executive Chair: Robert J. Howlett, University of Brighton, UK

KES2006 10th International Conference on Knowledge-Based & Intelligent Information & Engineering Systems, October 9-11 2006, Bournemouth. UK
Joint General Chairs: Bogdan Gabrys, University of Bournemouth and Robert J.Howlett, University of Brighton

KES'2005 9th International Conference on Knowledge-Based & Intelligent Information & Engineering Systems, September 14-16 2005, Melbourne, Australia
General Chair: Rajiv Khosla , La Trobe University, Melbourne, Australia
Executive Chair: Robert J.Howlett, University of Brighton, U.K.

IEEE Congress on Evolutionary Computation (CEC)
Most recent events:

2007 IEEE Congress on Evolutionary Computation (CEC 2007), September 25-28 2007, Singapore
2006 IEEE Congress on Evolutionary Computation (CEC 2006), July 16-21 2006, Vancouver , BC, Canada
2005 IEEE Congress on Evolutionary Computation (CEC 2005) Septembrie 2-5 2005, Edinburgh, Scotland, UK
Genetic and Evolutionary Computation Conference (GECCO)
Most recent events:
2007 Genetic and Evolutionary Computation Conference (GECCO-2007) July 7-11 2007 London, UK
2006 Genetic and Evolutionary Computation Conference (GECCO-2006) July 8-12 2006, Seattle, WA,USA

EHW BOOKS

- **Higuchi T., Liu Y., Yao X.,** (Eds)" *Evolvabe Hardware*", **Springer-Verlag, 2006**, http://www.springer.com/east/home/generic/search/results?SGWID=5-40109-22-46626015-0
- **Greenwood G. W., Tyrell A. M.,** " *Introduction to Evolvabe Hardware*", **Willey-VCH, 2006**, http://www.wileyvch.de/publish/en/books/bySubjectEE00/bySubSubjectEE10/0-471-71977-3/?sID=d05b
- **Thompson, A.,** *"Hardware Evolution: Automatic design of electronic circuits in reconfigurable hardware by artificial evolution"*, **Springer-Verlag, 1998**, http://www.cogs.susx.ac.uk/users/adrianth/ade.html
- **Zebulum et al,** *"Evolutionary Electronics: Automatic Design of Electronic Circuits and Systems by Genetic Algorithms"*, **CRC Press, 2001** http://www.crcpress.com/shopping_cart/products/product_detail.asp?sku=0865
- **Sekanina, L.,** *"Evolvable Components From Theory to Hardware Implementations"*, **Springer, 2003**, http://www.springer.de/cgi-bin/search_book.pl?isbn=3-540-40377-9&cookie=done
- **Negoita et al, "***Computational Intelligence. Engineering of Hybrid Systems*", **Springer, 2004 , ISBN 3-540-23219-2**
- **John Koza, "*Genetic Programming: On the Programming of Computers by Means of Natural Selection*" published by The MIT Press , 1992;**
- **John Koza, "*Genetic Programming II: Automatic Discovery of Reusable Programs*" published by The MIT Press, 1994.**
- **John Koza, "*Genetic Programming III: Darwinian Invention and Problem Solving*" published by Morgan Kaufmann Publishers, 1999.**
- **John Koza, "*Genetic Programming IV: Routine Human-Competitive Machine Intelligence*" (with Martin A. Keane, Matthew J. Streeter, William Mydlowec, Jessen Yu, and Guido Lanza) published by Kluwer Academic Publishers, 2003.**
- **Goldberg, D.**, *"Genetic Algorithms in Search, Optimization and Machine Learning"*, **Addison-Wesley Publishing Company, Inc., Reading, Massachusetts, 1989**.
- **Holland, J.**, *"Adaptation in Natural and Artificial Systems"*, **University of Michingan Press, Ann Arbor, EUA, 1975**.
- **Higuchi, T. ,** *"Evolvable Hardware and its Applications"*, **chapter in "Computational Intelligence The Expert Speak" by Fogel and Robinson, IEEE Press, 2003.**

- **Miller, J. F., Thomson, P., and Fogarty, T.,** *"Designing Electronic Circuits Using Evolutionary Algorithms. Arithmetic Circuits: A Case Study"*, **chapter 6 in Genetic Algorithms Recent Advancements and Industrial Applications. Editors: D. Quagliarella, J. Periaux, C. Poloni and G. Winter, Wiley, 1997**.

EHW JOURNALS

- **Genetic Programming and Evolvable Machines:** http://www.kluweronline.com/issn/1389-2576
- **International Journal of Knowledge-Based and Intelligent Engineering Systems:** http://www.iospress.nl/loadtop/load.php?isbn=13272314
- **IEEE Transactions on Evolutionary Computation:** http://www.ieee-nns.org/
- **Evolutionary Computation Journal (MIT Press) :** http://www.mitpress.mit.edu/EVCO/

Special issues on EHW in following journals:

- **International Journal of SMART ENGINEERING SYSTEM DESIGN**

- Soft Computing Journal, Special Issue on Evolvable Hardware **Adrian Stoica** , http://www.springer.de
- IEE Proceedings Computer-Digital Techniques Special Issue on Evolvable Hardware, **Andy Tyrrell** , Ed http://www.iee.org/Publish/Journals/ ProfJourn/ Proc/CDT/evolvable_hardware.pdf
- International Journal of Knowledge-Based and Intelligent Engineering Systems Special Issue on Evolvable Hardware, **Mircea Negoita** and **Tughrul Arslan** , Eds http://www.iospress.nl/loadtop/load.php?isbn=13272314

AHS/EHW RESEARCH GROUPS

NASA JPL Evolvable Hardware Laboratory (http://ehw.jpl.nasa.gov/)

Focussed to develop and demonstrate self-reconfigurable circuits that evolve directly in hardware on a VLSI chip. Self-reconfiguration is directed by this group to endow devices with the flexibility of in-situ, during the mission, adaptation to unforeseen conditions and with enhanced fault-tolerance. Their evolved circuits perform complex signal processing functions, such as adaptive filtering and randomization. JPL EHW group was successful in developing tools and approaches, and in establishing a technology base that others can use.

This group was involved in projects as follows: **EHW** for sensors; **EC** techniques for space science; **EHW** for adaptive computing; polymorphic electronics; humanoid robots for intelligent handling and assembly tasks.

The group leader is *dr. Adrian Stoica*, member of IEEE Task Force on **EHW**.

EHW Group at the Intelligent Systems Research Group of the University of York, UK (http://www.bioinspired.com/)

The research interest of this group are oriented in the following directions: biologically inspired architectures: embryonics, immunotronics, evolvable architectures; fault tolerant computer systems; distributed processor systems; models for concurrent systems and application of parallel systems.

This group is involved in the development of novel biologically-inspired computing machines and programs inspired by Nature and endowed with capabilities such as adaptation, evolution, growth, healing, replication, and learning; the characterisation and understanding of biological and biomedical signals; the exploitation of multiagent mechanisms in control systems.

The group leader is *prof. Andy Tyrell*, Chair of IEEE Task Force on **EHW**.

EHW Group at the Advanced Semiconductor Research Center of National Institute of Advanced Industrial Science and Technology (AIST) , Japan (http://unit.aist.go.jp/asrc/asrc-5/index_en.html)

The research area of this group is involved in the following directions: digital **EHW**, analog **EHW**, optical **EHW** and remote support system for handicapped people.

This group is a world leader department in development of a large area of **AHS/EHW** real-world applications and even of specialized **EHW** chips for the electronics industry.

Achievements in digital **EHW** applications are as follows: character recognition; autonomous mobile robots; self-reconfigurable **NN** chip; EMG-prosthetic hand; data compression for graphic arts.

Analog **EHW** applications were relevant by: an analog **EHW** chip for cellular phones; evolvable microwave circuits; a very useful evolutionary circuit design **EHW** relied method for adaptive clock timing adjustment of **VLSI** circuits; high – speed data transfer circuits with adaptive adjustments.

The group leader is *dr. Tetsuya Higuchi,* member of IEEE Task Force on **EHW**.

EHW Group at Department of Computer Systems, The Faculty of Information Technology (FIT), Brno University of Technology, Czech Republic (http://www.fit.vutbr.cz/research/groups/ehw/)

This group is involved both in *evolutionary circuit design* and in EHW.

The research and achievements of this team are to be classified in the following directions: evolvable hardware in a single FPGA; development platforms for EHW application; polymorphic electronics for obtaining circuits that are able to perform more functions depending on the environment in which they operate; theory and applications of evolvable components; evolutionary design of innovative, testable, and fault-tolerant circuits; theory and applications of virtual reconfigurable circuits; real-world applications of evolvable hardware.

The group leader is *prof. Lukas Sekanina*, member of IEEE Task Force on EHW.

The Reconfigurable and Embedded Digital Systems Institute (REDS) at The university of Applied Sciences (HEIG-VD), Switzerland (http://reds.heig-vd.ch/)

The **REDS** institute is involved in different activities regarding the area of complexity and the particularity of modern digital reconfigurable and embedded systems: applied research and development; technology transfer and technological support to the **SME**'s disposal for their industrial realizations; training of Bachelor Engineers; training and postgraduate training for industrial practitioners. Some of **REDS** technical profile covers typical **AHS** fields as follows: design of **S***ystems* **o***n* **P***rogrammable* **C***hip* (**SoPC**), of pervasive systems, and of dynamically reconfigurable systems; design of bio-inspired systems, adaptable to the environment and the users; design of bio-inspired reconfigurable hardware systems: neural hardware, evolvable hardware, and self-repairinghardware.

The institute runs complex national and international **AHS** R&D co-operation projects with partners from academic and/or industrial environment, as for example: PERPLEXUS project - in the field of bio-inspired pervasive reconfigurable systems; COCH project - in the field of the bio-inspired ecology.

PERPLEXUS is an European project focused on development of a scalable hardware platform made of custom reconfigurable devices with bio-inspired capabilities. The platform is aimed to be a simulation tool for large-scale complex systems , actually for three distinct applications: neurobiological modeling, culture dissemination modeling and cooperative robotics (Sancehez et al 2007).

PERPLEXUS 's *Ubichip* is a custom reconfigurable electronic device aimed to be capable of implementing bio-inspired circuits featuring growth, learning and evolution, namely evolvable and developmental cellular and neural systems reliant on routing, self-reconfiguration and a neural-friendly logic cell's architecture (Upegui et al 2007).

A contact person is prof. Eduardo Snanchez, member of IEEE Task Force on EHW.

Chapter 3
Bio-Inspired Analogue and Digital Circuits and Their Applications

3.1 Introduction

E*volvable* **H***ardware* consists of applications of **E***volutionary* **C***omputation* in the design of electronic circuits (Zebulum et al. 2002). It can be applied in many different areas such as automatic synthesis of analogue and digital circuits, developing of adaptive systems, building of fault tolerant systems, yielding low-power circuits, improving parameters obtained in VLSI circuits design (Gordon et al. 2002).

Automatic synthesis of electronic circuits is one of the most preferred applications of Evolvable Hardware. Many analogue circuits were studied as amplifiers (Zebulum et al. 1998), filters (Murakawa et al. 2001, Takahashi et al. 2003), analog multipliers (Zebulum et al. 2002), but also digital blocks, including logic gates (Gajda et al. 2007, Slowik et al. 2006), arithmetic circuits as adders (Miller et al. 1997, Aoki et al. 1999) or digital multipliers (Vassilev et al. 2000).

In the category of **E***volutionary Algorithms*, the **G***enetic Algorithms* are the most used in Evolvable hardware applications.

The Genetic Algorithms are based on the Darwinian principle of natural selection and they simulate the biological evolution. They propose a population of potential solutions starting with an initially randomly generated population of individuals.

Individuals are described by the information contained in chromosomes. Instead of the DNA present in nature, in EHW applications a binary string having a varied length is used, depending on the number of parameters to be optimized.

The fitness is a quantity describing the performance of an individual and it is defined in such a way that it can be optimized in the designing process.

In order to optimize the individual's fitness, GA performs then operations like selection, crossover and mutation over the individuals, similar with the natural operations.

For example, the crossover operation changes the genetic material between two parents, producing two new individuals or offspring. The mutation operator randomly changes the position of genes in a string. We are talking about a search process which will stop depending on a stop criteria and which consists of a generation of successive populations.

M.G. Negoita, S. Hintea: Bio-Inspired Tech. for the Hardware of Adaptive Sys., SCI 179, pp. 83–168.
springerlink.com

A very promising field is to use genetic algorithms in tuning reconfigurable analogue circuits applied in mobile communications (Murakawa et al. 1998; Hintea et al. 2007) as a part of more complex adaptive systems. In this case the evolutionary methods are used to correct the variations of electronic components during the fabrication process, because it is known that the fabricated values differ from the original designed ones. There are proposed capacitors or resistances networks capable to be switched in an established range with the smallest possible step. The values are digitally controlled using genetic algorithms which are able to react depending on the real values of the passive components which could be different in each case of designed chip.

These techniques are very useful in tuning intermediate frequency filters for mobile telecommunications devices but also to design programmable Variable Gain Amplifiers, as we show in this chapter. This method could lead to less area and power consumes and increased yields, too.

These analogue applications could be extended to the field of biomedical devices. In the case of auditory prosthesis the control of the signal amplitude transmitted to the ear is an important task and it could be achieved by digitally controlled VGA's. One Automatic Gain Control (AGC) is able to maintain a constant signal level, when the environment could dramatically influence the sound level received by the human subject. In all these situations, a genetic algorithm could change the response of analogue circuits and adapt their characteristics to a different situation.

In the field of implantable auditory prosthesis, the adaptive system plays a very important role. The use of evolutionary methods could be also used to design other sorts of implantable stimulators. Some systems have been developed such as the one described by (Kajitani et al. 1998) in order to control a prosthetic hand in a user's arm. This implementation consists of a GA base hardware solution, reconfigurable logic, a chromosome memory and a CPU core.

3.2 Genetic Algorithms for Analogue Circuits Design

3.2.1 GA as Tools to Design Analogue Circuits

The genetic algorithm is an unconventional technique used to find better solutions for optimization problems in the design of analogue circuits (Azizi 2002). Using GA is interesting especially because they allow the automatic design of the circuits intended to be used with different purposes as industrial, telecommunications, aerospace and medical applications.

Beside the automation of the designing processes, EHW methods allow the achievement of some circuits having improved performances in terms of: (Zebulum et al. 2002):

- Fault tolerant design;
- Yielding low-power circuits;

- Circuits for extreme environment conditions;
- Synthesis of fuzzy controllers.

During the last decade important steps were made in the field of using techniques from evolutionary computation in designing analogue circuits. These include the use of GA's to design different circuits as:

- Various analogue filter design problems, including select filter topology or component sizes;
- Design operational amplifiers using a small set of topologies;
- LNA and other amplifiers design;
- The topology of passive linear circuits composed by capacitors, resistors and inductors;
- Temperature sensing circuit;
- Voltage reference;
- Computational circuits;
- Robots controllers;
- Aerospace applications.

Design optimization of an electronic circuit is a technique used to find the design parameter values (length and width of MOS transistors, bias current, capacitor values etc) in such a way that the final circuit performances (dc gain, gain-bandwidth, slew rate, phase margin etc.) meet as close as possible the design requirements (Oltean and Hintea 2008).

GAs used for the design of analogue circuits usually follow the basic GA flow or other more complex GA which are all extensions of this basic flow. The first step consists of creating the initial random population of individuals. This initial population is usually generated randomly, but sometimes it could be populated with solutions that are known to have characteristics close to the desired solution. These solutions are also chosen by taking into account the previous experience of the designer, which sometimes can be a disadvantage in finding new and unconventionally solutions.

The next step is to transform the string representing the chromosomes into a format recognizable by a circuit simulator such as SPICE. The circuit simulator will simulate each solution, and then the fitness function is applied on the simulator output data to produce a fitness measure for each solution. This is the method called *Extrinsic evolution*. In the case of *Intrinsic evolution* each resulted candidate circuit corresponding to every population is evaluated in a physical reconfigurable circuit (Sekanina 2006).

As in the general GA, the steps of selection, such as crossover, mutation, and the re-evaluation of fitness measures are applied iteratively until a solution that meets the specifications is found. We can see the general flow in figure 3.1 (Azizi 2002).

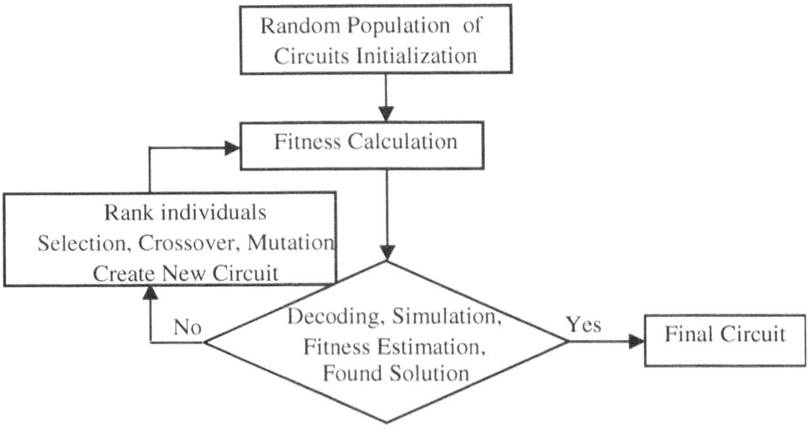

Fig. 3.1 GA flow for Analogue Circuits Synthesis

3.2.2 *Overview of the Genetic Algorithm*

Central to all genetic algorithms is the concept of the chromosome. The chromosome contains all information necessary to describe an individual. In nature, chromosomes are composed of DNA. In a computer, a long binary or character string is used. Chromosomes are composed of genes for the various characteristics to be optimized and can be any length depending on the number of parameters to be optimized. Basically, in a genetic algorithm each chromosome is potentially a solution of the optimization problem. Encoding defines the way genes are stored in the chromosome and translated to actual problem parameters.

A generic chromosome employed in genetic algorithms is shown in figure 3.2 where each gene represent a design parameter.

Fig. 3.2 Format of a generic chromosome

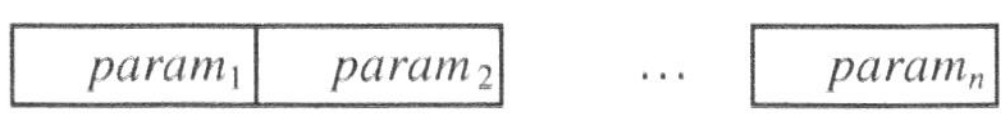

The "alphabet" used in the representation of genes can, in theory, be any finite alphabet. Thus, rather than use the binary alphabet of 1 and 0, one can use an alphabet containing more characters and numbers. Because the design parameters are real variables we choose real numbers to compose our alphabet. This way our population individuals are real valued vectors, rather than bit strings, thus simplifying the development of GA implementation.

The underlying procedure of a GA is shown in Figure 3.3.

P*opulation* **I***nitialization* is commonly done by seeding the population with random values, with a uniform probability. It is sometimes feasible to seed the population with "promising" values that are known to be in the hyperspace region relatively close to the optimum. Each individual in the selection pool receives a

Fig. 3.3 GA procedure

1. *Population initialization*
2. *Fitness assignment*
3. *Selection*
4. *Recombination*
5. *Mutation*
6. *Reinsertion*
7. *Loop to step 2 until optimal solution is found*

reproduction probability depending on the own objective value and the objective value of all other individuals in the selection pool. This **F***itness* is used for the actual selection step afterwards. The implementation uses rank-based fitness assignment with linear ranking (Pohlheim 2005).

Consider *Nind* the number of individuals in the population, *Pos* the position of an individual in this population (least fit individual has *Pos*=1, the fittest individual *Pos=Nind*) and *SP* the selective pressure. For example, in the case of a minimization-type optimization problem first position is allocated to the individual with highest value of the objective function. The fitness value for an individual is calculated as:

$$Fitness(Pos) = 2 - SP + 2(SP - 1)\frac{Pos - 1}{Nind - 1} \quad (3.1)$$

Linear ranking allows values of selective pressure in [1.0, 2.0].

The roulette-wheel **S***election* is a stochastic algorithm and involves the following technique. For each individual a selection probability is computed as:

$$Selection\_probability(i) = \frac{Fitness(i)}{\sum_{i=1}^{N} Fitness(i)} \quad (3.2)$$

The individuals are mapped to contiguous segments of a line, such that each individual's segment is equal in size to its selection probability. A uniformly distributed random number is generated and the individual whose segment spans the random number is selected. The process is repeated until the desired number of individuals is obtained (called mating population). This technique is analogous to a roulette wheel with each slice proportional in size to the fitness (Oltean and Hintea 2008).

R*ecombination* produces new individuals in combining the information contained in two or more parents (parents - mating population). This is done by combining the variable values of the parents. Depending on the representation of the variables different methods must be used. For our real valued variables the intermediate recombination method was chosen. Offspring are produced according to the rule (Pohlheim 2005):

$$Var_j^O = a_j Var_j^{P1} + (1 - a_j) Var_j^{P2}, \quad j = 1, 2, ..., Nvar \quad (3.3)$$

where Var_j^O represent the jth variable of the offspring, Var_j^{P1} represent the jth variable of the first parent, while Var_j^{P1} represent the jth variable of the second parent. The scaling factor a_j is chosen uniformly at random over an interval $[-d, 1+d]$ for each variable anew.

Intermediate recombination is capable of producing any point within a hypercube slightly larger than that defined by the parents.

By **M***utation*, individuals are randomly altered. Mutation of real variables means that randomly created values are added to the variables with a low probability. Thus, the probability of mutating a variable (mutation rate) and the size of the changes for each mutated variable (mutation step) must be defined.

The probability of mutating a variable is inversely proportional to the number of variables (dimensions). The more dimensions one individual has, the smaller is the mutation probability. Different papers reported results for the optimal mutation rate. (Oltean 2008) writes that a mutation rate of $1/n$ (n: number of variables of an individual) produced good results for a wide variety of test functions. That means that per mutation only one variable per individual is changed/mutated. Thus, the mutation rate is independent of the size of the population.

The size of the mutation step is usually difficult to choose. The optimal step-size depends on the problem considered and may even vary during the optimization process. It is known, that small steps (small mutation steps) are often successful, especially when the individual is already well adapted. However, larger changes (large mutation steps) can, when successful, produce good results much quicker. Thus, a good mutation operator should often produce small step-sizes with a high probability and large step-sizes with a low probability.

Such an operator (Pohlheim 2005) could be considered for the algorithm:

$$\begin{aligned}
&Var_j^{Mut} = Var_j + s_j r_j a_j, \quad j = 1, 2, ..., Nvar \\
&s_j \in \{-1, +1\} \text{ uniform at random} \\
&r_j = r \cdot domain_i, \ r \text{ - the mutation range (standard 10\%)} \\
&a_j = 2^{-uk}, \ u \in [0, 1] \text{ uniform at random, } k \text{ - the mutation precision}
\end{aligned} \tag{3.4}$$

The range of mutation is given by the value of the parameter r and the domain of the variables. The parameter k (mutation precision) defines indirectly the minimal step-size possible and the distribution of mutation steps inside the mutation range. The smallest relative mutation step-size is 2^{-k}, the largest 2^0=1. By changing these parameters (r and k) very different search strategies can be defined.

The **R***einsertion* scheme produce as many offspring as parents and replace all parents by the offspring. Every individual lives one generation only.

3.2.3 Representation

The representation of a circuit is made with the help of a chromosome or genotype, which is a string of bytes, each byte being called gene. Each chromosome describes all the components of the circuit or the sub-circuit with the type, value and connection points.

The representation establishes the direct connection between the integer strings representing the chromosomes and the electronic circuit topology. The genes, which are the bits components of the whole string, describe the type, value and connecting points of the electronic component. This component could be a passive one, like resistors, capacitors or coils, or an active one, like transistors (Zebulum et al. 2002).

We describe such a simple example in figure 3.4, where a common source amplifier is built around a MOS transistor. We have three electronic components, each of them having a value and two points of connection.

In this example the chromosome consists in three genes: one of them describes the transistor: Gene1 = [0,-,(2,1,0)] where we have chosen the type '0' which means "transistor" , it has no value (-) and it is connected in three points: 2, 1 and 0.

Gene2 = [1,1k,(2,3)] where the type of component '1' means "resistor", the value is 1kΩ and it is connected between the points 2 and 3.

Gene3 = [2,1n,(2,0)] where the type of component '2' means "condenser", the values is 1nF and it is connected between the points 2 and 0.

Fig. 3.4 Analogue Circuit's Representation

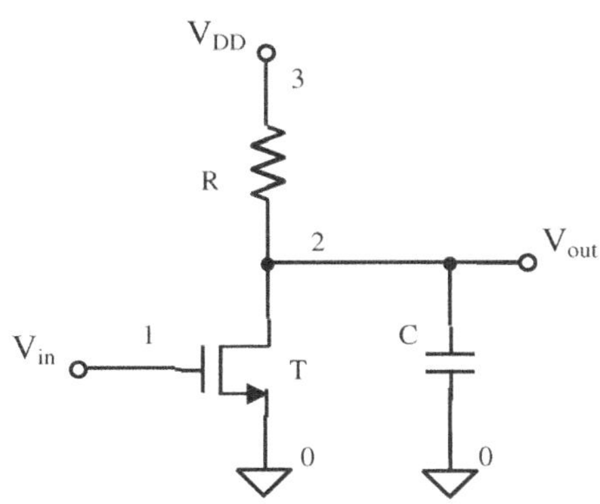

Choosing the number of connection points, as a parameter of the representation, is very important taking into account that if the number of points is too large many topologies that can't be simulated may appear, while if the number is too small an insufficient number of topologies will result (Zebulum et al. 2002).

The codification principle is described in figure 3.5 where *tt* encodes the component's type, *nnnn* node 1, *mmmm* node 2, while *vvvvvv* the component's value.

tt nnnn mmmmm vvvvvv

Fig. 3.5 The codification principle of a component

In what concerns the values of the resistances and condensers a convention that uses the 8 gene chromosome can be used (Azizi et al. 2002). Each gene represents a component with the help of two fields, from which one describes the decade of the component's value between 10^3 and 10^6 and the last 4 bites show which value from the 12 existent of the decade is selected for that particular component. The representation for the passive components is described in figure 3.6.

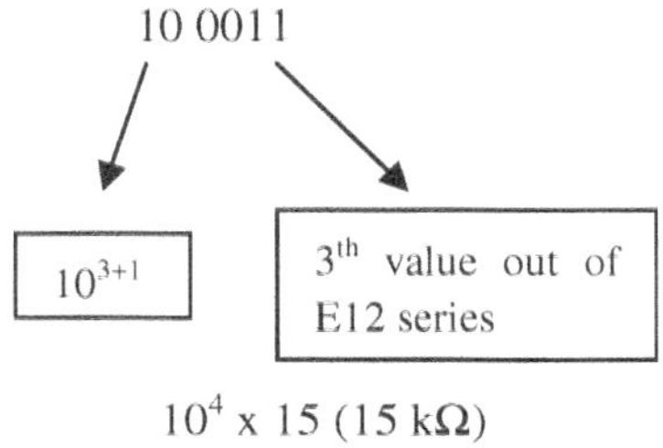

Fig. 3.6 The chromosomal representation of a passive circuit

In the case of the active components a description with 10 gene chromosomes is used, each gene representing one by one the width and length of the channel, the bias current, the value of the compensation capacity (Zebulum et al. 1998) as it can be seen in figure 3.7.

W_1 L_1 W_2 L_2 C_B I_B

Fig. 3.7 The chromosomal representation of an active circuit

The designing manner in which the topology is chosen at the beginning and GA is used only in choosing the value and in positioning the components is limited and it depends on the value of the initial scheme. It doesn't allow an extension towards schemes which weren't previously used and, this way, the more advantageous use of the allowed innovations of GA.

A suggestive application having good results was developed by (Zebulum et al. 1998): an AO with CMOS transistors in a standard Miller OTA configuration, having the simplified scheme in figure 3.8.

It is a scheme with fixed topology, in which the values of the components are chosen: the transistors' dimensions, the biased currents and the compensation condenser. These values compose the genotype positions of the chromosome.

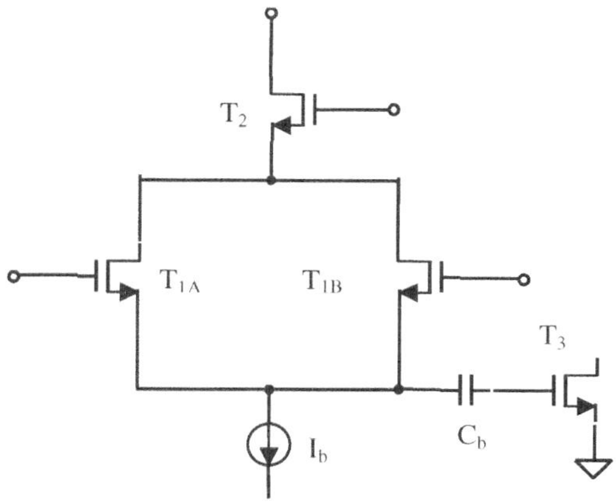

Fig. 3.8 OpAmp chromosomal representation

In order to achieve this optimization certain conditions are imposed to the upper values: the width and length of the channel are greater than the minimum values W_{min} and L_{min}. One hundred discrete values were taken between the minimum one and the minimum one plus 100 at intervals of 0.01 μF. The bias currents were chosen with values between 1.5 μA and 2.5 μA, also 100 values. The compensation capacity C_b is taken in the domain 0.1 pF and 10pF, 100 values at intervals of 0.1 pF.

An interesting aspect is that the circuits obtained by using GA fulfill some rules which are followed also in the traditional design. For example, the pairs of transistors are chosen having equal dimensions, high values for the W/L ratios of the transistors which determine a high gain or a high slew-rate.

The schemes obtained by applying GA are comparable in terms of performances to the ones designed in a classic manner, some of the characteristics being even better (for example the dissipated power is lower).

In the same domain, (Zebulum et al. 2002) presented an application of an analogical multiplier made in CMOS technology, an important part in many analogical systems, such as filters, support vector machines, mixers and modulators. The evolved CMOS multiplier uses 6 transistors, which is a more efficient result than the 19 transistors of the conventional circuit.

In this application the fitness is calculated as the sum of the squared deviations between the actual output and the target value. The circuit is obtained after only 200 generations, sampling 50 individuals with an average percentage error to the target of 2.43% and the highest error equal to 10.18% (Zebulum et al. 2002).

3.2.4 Analogue Applications with FPTA Cells

One intrinsic implementation of Evolvable Hardware was proposed by (Stoica et al. 2001) and uses field programmable transistor arrays (FPTA), a reconfigurable architecture at transistor level. They used a FPTA cell fabricated in 0.5 μm CMOS technology, every test chip containing two FPTA cells.

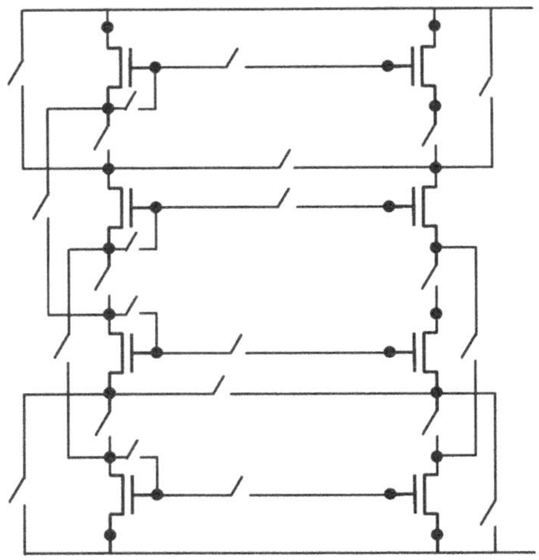

Fig. 3.9 FPTA cell schematic

One FPTA cell is described in figure 3.9, where we can see that every cell contains 8 transistors and 24 programmable switches. The circuit topology and the circuit response are determined by the status of programmable switches, which allow the interconnection of the array of transistors. The topology will be a function of switch states and will depend on a binary sequence, where the bits values will correspond to the switches states ("ON" or "OFF").

The 8 transistors, 4 NMOS and 4 PMOS, allow an important number of analogue applications and the most important fact is that this hardware platform provides a much higher processing speed than the simulated solutions made on computers.

A number of applications were designed on such a FPTA platform: amplifier, band pass filter with a passing band between 1 and 10 kHz and a circuit with Gaussian voltage-input current-output characteristics (Stoica et al. 2001).

The advantages of this sort of structure are:

- The great variety of analogue circuits schemes, from the simplest to the most complex;
- provide a hardware platform for intrinsic evolution algorithms;
- a much higher speed than the simulated extrinsic evolution solutions;
- easy conversion to VLSI CMOS technology.

3.2.5 Design Optimization of a CMOS Amplifier

3.2.5.1 Formulation of the Optimization Problem

The optimization algorithm begins with the formulation of the optimization objectives and optimization problem, followed by the initialization of the design parameters. During iterations an evaluation engine computes the actual circuit performances based on the actual design parameter values. If the objectives are

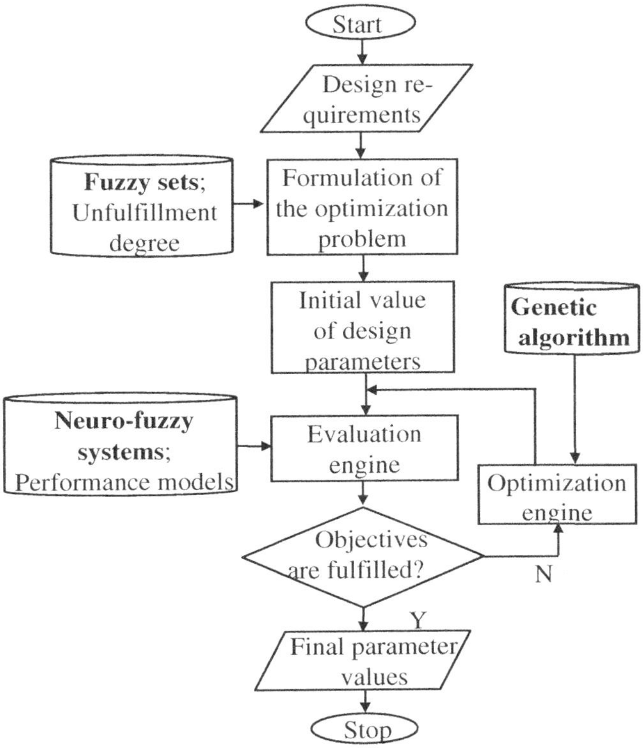

Fig. 3.10 CI-based optimization algorithm

fulfilled, the solution consists in the set (or sets – in the case of a real multiobjective optimization) of the actual design parameter values and the algorithm is stopped. If not, new design parameter values are to be computed by the optimization engine and the optimization loop is covered once again.

The novelty introduced by (Oltean and Hintea, 2008) is the utilization of different CI techniques in all phases of optimization algorithm, as it is shown in Fig. 3.10.

Fuzzy sets are used to define the objective function in formulation of the optimization problem. Neuro–fuzzy systems address the performance evaluation issue (evaluation engine). Finally, in the optimization engine, a genetic algorithm is responsible for the evolution of the population to finally produce the near optimum solution.

To solve a multiobjective optimization problem, as is the design optimization of analogue circuits, the optimization problem can be formulated as a single-objective optimization, where different performance objectives are combined to form a single scalar objective, and which produces one solution.

The utilization of fuzzy sets to define the objective functions is proposed in (Oltean 2005). By contrast with the existing approaches where membership degree represents the degree of fulfilment, in this approach the membership degree μ

represents the error degree in the fulfilment of the objective. A value μ=1 means the objective is not satisfied at all, while a value μ=0 means that the objective is fully satisfied. With this approach we know the range of possible values for objective functions as being [0, 1]. When the value of a certain objective function (unfulfilment degree - *UD*) is 0, we know that the corresponding requirement is fulfilled, no further effort being necessary to improve the associated performance.

For a single-objective optimization approach, the individual objective functions are combined into a cost function by means of a weighted sum. The formulation of the multiobjective optimization problem becomes:

$$\text{Find } x \text{ that minimizes } F(x) = \sum_{k=1}^{n} w_k \mu_k \left(f_k(x)\right) \tag{3.5}$$

where n is the number of requirements, f_k is the kth objective function, and w_k is the weight or relative preference associated with the kth objective function.

3.2.5.2 Evaluation Engine

Fuzzy systems are very useful to model circuit performances because they are considered universal approximators. (Oltean 2005) has been synthesized a method to build neuro-fuzzy models of circuit performances. These models are automatically built up based on input-output data sets, using ANFIS (Adaptive Neuro-Fuzzy Inference System) framework (Jang R J-S 1993) to develop first order Takagi-Sugeno fuzzy systems. ANFIS framework contains a six layer architecture for its artificial neural network trained by a mixture of back propagation and least mean square estimation learning algorithm (Jang 1993).

The parameter set (the combination of the parameter values) should be chosen to be representative for the performance to be modelled (covers thoroughly the parameter space and embeds all the specific characteristics of the performance). For each input vector (one combination of the parameter values), the performance value has to be found, in our case by SPICE simulation. Two data sets, a training set and a checking set are generated. A subtractive clustering procedure generates an initial first order Takagi-Sugeno fuzzy system. Next the fuzzy set is trained and validated using previously generated data sets.

3.2.5.3 Optimization Engine

The heart of the whole algorithm is the optimization engine. A genetic algorithm (GA) is responsible for the exploration of solution space in quest of the optimal solution. Generally, the best individuals of any population tend to reproduce and survive, thus improving successive generations (Ren 2007). However inferior individuals can, by chance, survive and reproduce. In this case, the individuals consist of different versions (same topology, but different parameter values) which can evolve until a solution is reached (in terms of requirements accomplishment).

3.2.5.4 Design Optimization of a CMOS Amplifier

The proposed CI-based design optimization algorithm is implemented in the Matlab environment. The algorithm is used for the design optimization of a simple operational transconductance amplifier shown in figure 3.11. The design parameters of the circuit are the dimensions of the transistors $(W/L)_1=(W/L)_2$, $(W/L)_3=(W/L)_4$, $(W/L)_5=(W/L)_6$ and bias current I_b. As performances, the important ones were considered: voltage gain A_{vo}, unity gain bandwidth GBW, and common mode rejection ratio $CMRR$.

Fig. 3.11 Simple operational transconductance amplifier

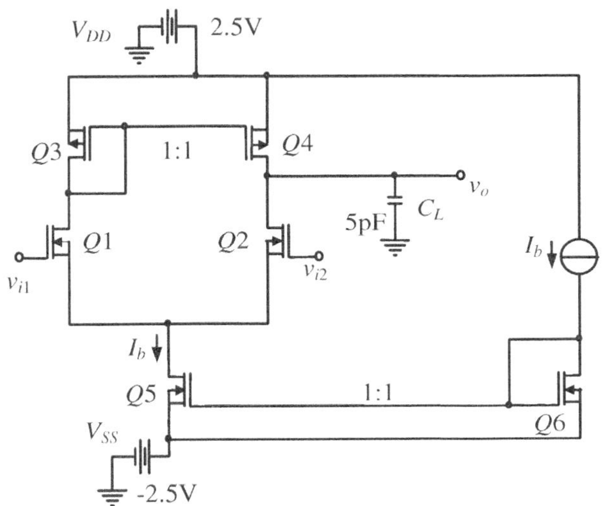

Table 3.1 Requirements and performances in four optimization runs

Circuit functions	Requirements	Performances			
		Run1	*run2*	*run3*	*run4*
A_{v0}	≥ 50	52.17	52.01	51.83	51.97
GBW [kHz]	$\geq 4\,600$	4 601	4 620	4612	4608
$CMRR$	$\geq 1\,000\,000$	1 011 413	1 008 935	1 004 844	1 003 761
Number of iterations		94	90	87	63

The design optimization is illustrated here for the set of requirements presented in table 3.1, for equal weighted objective functions. The optimization was run several times, for a population of 100 individuals. The algorithm proved to be a robust one, always a solution being found that fulfils all the requirements. Different numbers of iterations are necessary to search for the optimum solution depending on the initial population and on the evolution process. Table 3.1 gives also the

Table 3.2 Solutions in four optimization runs

Design parameter	run1	run2	run3	Run4
$(W/L)_{12}$	39.28	39.73	39.33	39.07
$(W/L)_{34}$	4.00	3.99	3.98	3.99
$(W/L)_{56}$	6.90	6.86	7.21	7.23
I_b [μA]	96.33	97.22	94.74	94.00

final performances of the circuit after four different optimization runs, while table 3.2 shows the solutions (values of the design parameters). The solutions appear to be slightly different from each other. At a closer look we can see that the values of the design parameters are calculated with two decimals. In practical implementations these values should be rounded toward some discrete values, so the resultant solutions are in fact small variations around one solution – these solutions are near optimum solution.

For the first optimization run (*run1*), the dynamic behaviour of the algorithm is presented in Figure 3.12.

In the first iterations (up to 10), due to a high diversity of individuals, a new population does not contain always a fittest individual than in the previous population (minimum value of the cost function increases). On the other hand, the population as a whole is improved continuously, the average value of the cost function on the entire population decreasing in time. As the population improves during evolution, all individuals move toward the optimal solution, decreasing both minimum and average values of cost function.

The evolution of all performances during optimization is presented in Figure 3.13. The optimization effort is mainly spend to continuously increase *GBW*, while keeping the values for A_{vo} and *CMRR* large enough to satisfy the requirements (see Table 3.1).

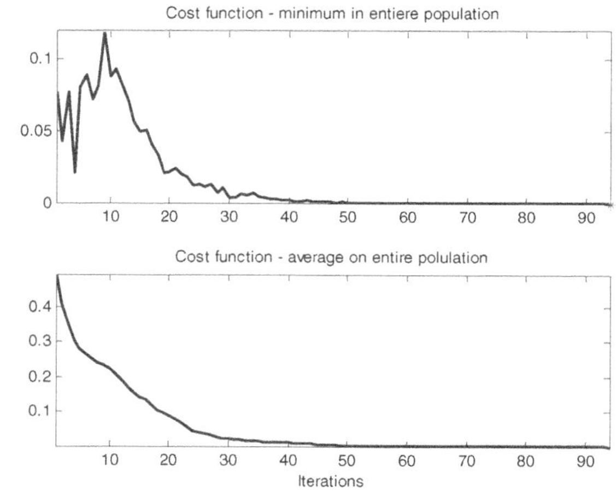

Fig. 3.12 Minimum and average cst function evolution for *run1*

Fig. 3.13 Dynamic behavior of performances in *run1*

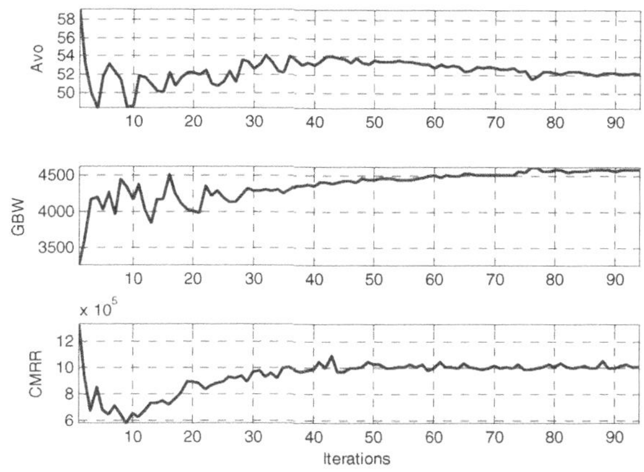

3.2.6 Evolving Software Models of Analogue Circuits

Some evolutionary methods used for design circuits are based on software models simulated by different simulators and were proposed in order to obtain better analogue circuits. The basic idea is the use of an automaton called circuit-constructing robot or cc-bot, dedicated to automatically design circuits. The language that programs the cc-bot contains component-placing instructions (Lohn et al. 1999).

A template circuit within the cc-bot builds the circuit. This template has one input and one output terminal as described in figure 3.14. Here the circuit's output voltage is an ideal voltage source V_s connected to ground and to a source resistor R_s. The output voltage is delivered across a load resistor R_l. One important aspect is that the size of the circuit can be evolved because the lists of cc-bot instructions have variable-length (Lohn et al. 1999).

Inside this template circuit the set of cc-bot instructions are able to lead to a valid electrical circuit and as a consequence the genetic algorithm will deliver valid circuit graphs.

The effect of every cc-bot instruction is the placement of a circuit component and the movement of the cc-bot. There are five basic instruction types as follows: x-move-to-new, x-cast-to-previous, x-cast-to-ground, x-cast-input, x-cast-to-output, where x can be replaced by R (resistor), C (capacitor), L (inductor), or transistor configuration (Lohn et al. 2000).

Cc-bot instructions are represented by up to four byte codes. So far, for instructions that take a component value as an argument, the first byte is the instruction, and the next three represent the component value (resistance, capacitance, and inductance values). For transistors, component values are not needed.

An overview of the evaluation process is depicted in Figure 3.15. The byte codes strings are transformed in a SPICE netlist representation. The netlist is processed by SPICE which delivers an output used to compute fitness for the individual.

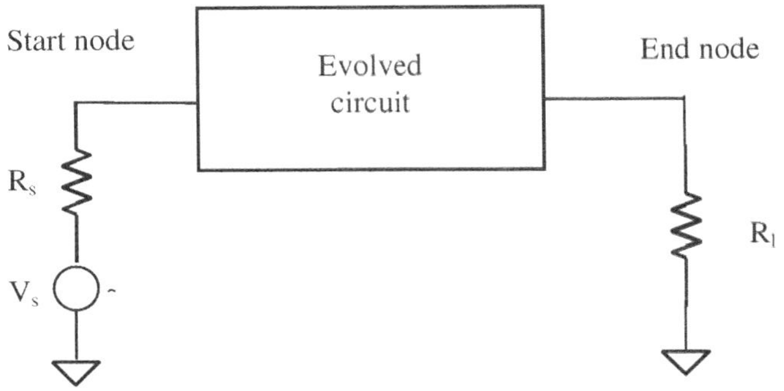

Fig. 3.14 The evolved circuit location

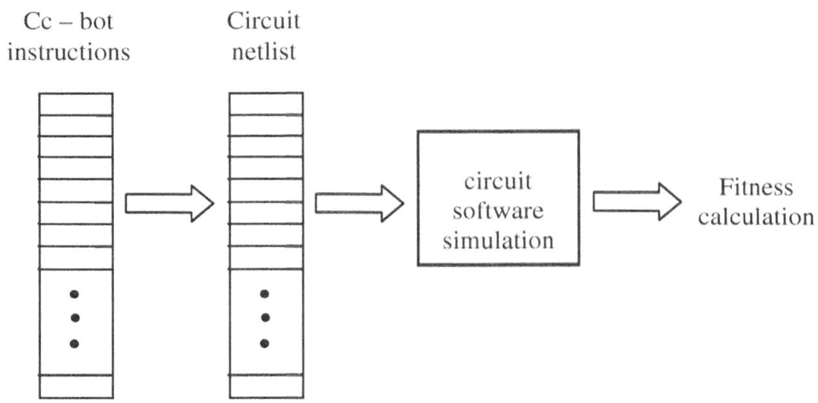

Fig. 3.15 Circuit evaluation process

The method was applied over some basic filters and amplifier configurations. The filters are RC or LC low pass filters, consisting of passive components (resistances, capacitances and coils). The simplest is a stethoscope RC low pass filter and the most complex, a fifth order Butterworth LC low pass filter. The size of populations varied from 3,000 to 18,000 (Lohn et al. 1999).

Two inverting amplifiers were also designed with the help of GA. The components are bipolar transistors, resistances and capacitors.

In the first case, the gain was set between 100 and 120 dB and the best performance was found in the generation 4866 and had a gain of 75 dB. The amplifier had a bandwidth of 7.59 kHz, a dc bias of 3.64 V and a power dissipation of 0.82 W. For the second amplifier the best performance was found in the generation 3635 and had a gain of 85 dB. For the chosen circuit the bandwidth was 282.9 kHz, a dc bias of 5.44 V and a power dissipation of 8.17 W. This method could be extended to many other circuits but the main limitation is the restriction on circuit topologies.

Another idea proposed by authors is the implementation of a parallel GA over a network of UNIX-based workstations. This could be useful because the processing time due to the simulation time could have very high values. But also we can say that this solution shows us an important drawback of those software simulation methods: if we need so many computers to design some small and simple electronic circuits it is difficult to replace the conventional methods in this way. And also this makes clearer the fact that hardware implementation of GA could be a better solution to design electronic circuits.

One other limitation is also the fact that the fitness function is calculated taking into account a single parameter. Amplifiers and filters have a whole set of parameters which are design requirements. In this case a multi-objective fitness function that accounts for each required parameter is an important improvement to evolve circuits like that.

3.3 Evolutionary Design of Digital Circuits

3.3.1 Combinational Logic Circuits Evolutionary Design

The principal task in implementing logic functions with logic gates consists in decreasing as much as possible the circuit's complexity, meaning to reduce the total number of gates and also the number of inputs of each gate. This leads to reductions in terms of power consumption and area size and as a consequence, the production costs. This could be achieved by using optimization tools such as evolutionary method, and they could obtain better designs from the point of view of number of gates and number of transistors. Reducing the number of gates is important especially when we use programmable logic devices to implement the solution. Reducing the number of transistors is useful when the implementation is made directly on the silicon.

The human designer uses traditional tools like Karnaugh map which is a graphical representation of logic functions or a tabular technique like Quine – McCluskey method. In order to obtain the best results the designer must be experienced and must respect some design rules.

One disadvantage when human designers use conventional methods is that Karnaugh diagrams minimizations lead to the utilization of NAND, NOR, NOT gates without the possibility to implement circuits with XOR gates. And it is known that many implementations could be more efficient with the help of XOR gates, which are fast and simple cells in different MOS technologies.

In the case of combinational logic circuits there are two optimization criteria: based on the gate count and based on the transistor count (Slowik and Bialko 2006). Both criteria are important but they depend also on the physical implementation. When they use the direct implementation on silicon, the minimization of the number of transistors is crucial. But when the desired function will be realized on FPGA circuits or other programmable hardware devices, the minimization of the number of transistors is more important (Slowik and Bialko 2006).

Evolvable Hardware is an alternate solution capable to offer better implementations in respect to the circuit size, speed and power consumption.

The first Designer Genetic Algorithms (DGA) was introduced by (Louis 1993) as an application of evolutionary design in digital circuits. Another contribution was the work of (Thompson 1996) where a new evolutionary method was proposed to design a tone discriminator circuit without input clock. (Miller et al. 1997) presented the solution of arithmetic circuits design with the help of genetic algorithms, which range from a simple one bit full adder to a 3-bit multiplier.

In this field of arithmetic circuits design important contributions were made by (Coello et al. 1997), (Vassilev et al. 2000) and (Gajda and Sekanina 2007). All these design methods based on genetic algorithms could be of great importance for automated design of digital circuits, but also when we need to implement adaptive solutions for different applications, meaning circuits capable of reacting to the modifying entries. One of the targets which is hard to be reached is to optimize not only the number of gates or transistors, but also the power dissipation and delay time of the obtained circuits.

One approach is based on a structure consisting of NOR/NAND gates (Gajda and Sekanina 2007) as it is described in figure 3.16.

Fig. 3. 16 Gate level implementation. a) NAND/NOR gate; b) NOR/NAND gate.

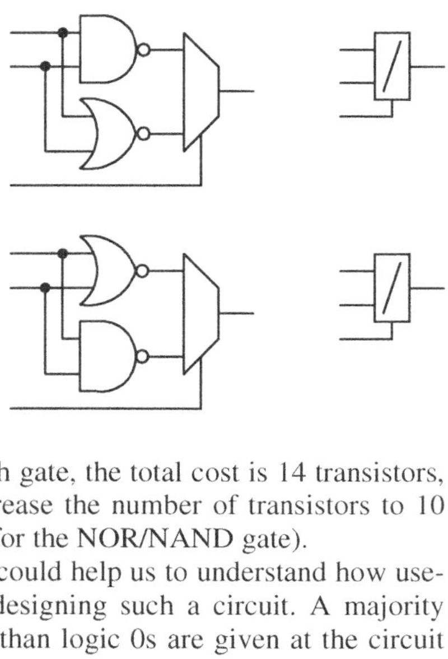

If we consider the implementation for each gate, the total cost is 14 transistors, other implementations being capable to decrease the number of transistors to 10 (in the case of the NAND/NOR gate) and 8 (for the NOR/NAND gate).

The majority circuit is an example which could help us to understand how useful an evolutionary algorithm could be in designing such a circuit. A majority circuit returns logic 1 only if more logic 1s than logic 0s are given at the circuit input. One conventional implementation is to use only two–input AND gates and two–input OR gates.

The results were obtained according to a sorting network-based implementation. The 5-input majority circuit consists of 10 such gates (i.e. 60 transistors (4 for the NAND gate, 4 for the NOR gate and 6 for the MUX) (figure 3.17). The other application, the 7-input majority circuit, consists of 20 such gates (fig. 3.18).

Fig. 3.17 Evolved 5-input majority circuit

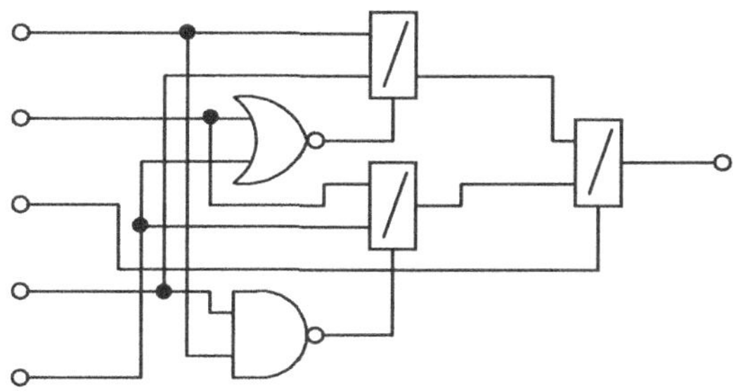

Fig. 3.18 Evolved 7-input majority circuit

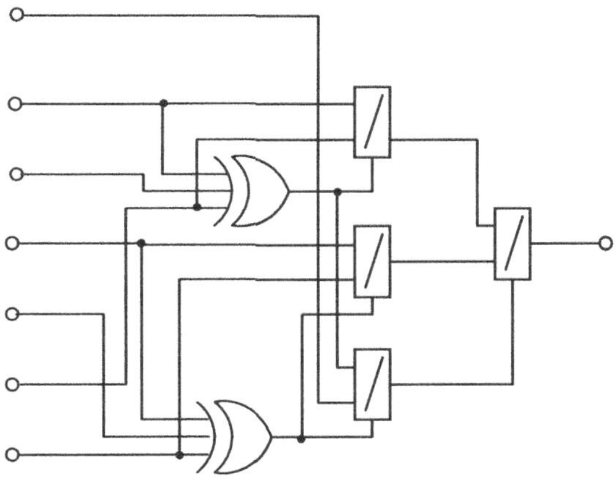

The evolved solutions use two unconventional gates and in most of the cases they contain fewer transistors than the same applications designed at transistor-level with the help of conventional techniques.

The time needed for a single run of EA varies in a range between 30 sec. and 198 sec. for the studied circuits (1 and 2-bit adder, 7-input majority circuit) using a standard PC equipped at the level of year 2007.

The method is optimized with regard of the gate level delay. But the evolved solution is not optimized for the transistor level and also, as in many other similar studies, the power dissipation, placement and routing aspects are not considered. Even the circuits are not evolved directly at the transistor level, however the implementation cost of candidate circuit was evaluated at the transistor level (Gajda and Sekanina 2007).

Another evolutionary method was introduced by (Slowik and Bialko 2006) with the name **M***ulti*-**L***ayer* **C***hromosome* **E***volutionary* **A***lgorithm* – **T***ransistor* **C***ount* (MLCEA -TC). This method is a modification of MLCEA introduced earlier by the same authors. This method consists in creating of an initial population with the help of a pattern of the designed circuit. The circuit is divided in sub-blocks which are coded in a multiple-layer chromosome. The structure consists of n elements, each having m features located in a single column as described in figure 3.19.

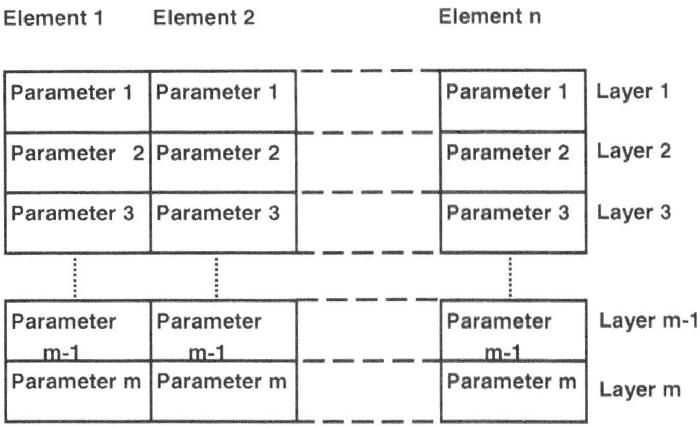

Fig. 3.19 Structure of multilayer chromosome

The main characteristic is the optimization of digital circuits with respect to the transistor count. There are used the following sort of gates: NOT, NOR, XOR, NAND, DC (direct connection of the gate input with its output).

Most of the evolutionary techniques applied in combinational circuits design optimize the designed circuit with respect to gates count. This is recommended especially when the implementation is made with the help of programmable devices, e.g. FPGA circuits, because these devices contain logic gates.

This method shows that it is possible to optimize the combinational logic circuits with respect to the transistors count. This is very useful when the design is directly implemented on silicon, because in this case the circuit size, power dissipation and the production costs are decreasing with the number of transistors (Slowik and Bialko 2006). It is true that the minimization of the number of transistors does not lead automatically to the minimization of the gate number.

3.3.2 Conventional Design Techniques for Arithmetic Adders and Multipliers

3.3.2.1 One Bit Full Adders

The adder is the central circuit of any arithmetic system, mainly because each operation can be reduced to a sum of additions. Any N-input adder consists of several 1-bit adders, connected in different ways.

One cell called the 1-bit full adder provides a sum of the 2 bits received at its input, while also taking into account a Carry input bit, which is generated by the previous stage. The adder having three inputs and two outputs drawn in figure 3.20 is called the 1-bit full adder. Here the C_i signal is the Carry bit received from the previous stage, while C_o is the output Carry provided for the next addition cell. The truth table is given in table 3.3.

Table 3.3 Truth table of a 1-bit full adder

Input a	b	C_i	Output S	C_o
0	0	0	0	0
0	0	1	1	0
0	1	0	1	0
0	1	1	0	1
1	0	0	1	0
1	0	1	0	1
1	1	0	0	1
1	1	1	1	1

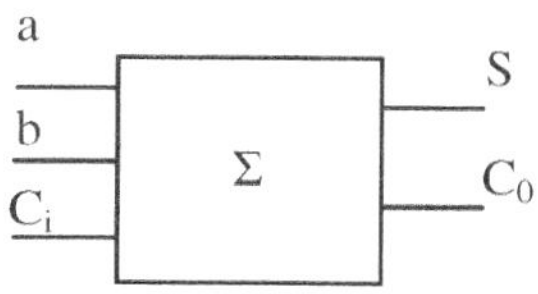

Fig. 3.20 Full adder logic symbol

After minimization, the expression of C_o becomes:

$$C_o = ab + bC_i + aC_i \tag{3.6}$$

The output S cannot be minimized, but it can be written using the XOR operator as follows:

$$S = a \oplus b \oplus C_i \tag{3.7}$$

The gate implementation of relations (3.6) and (3.7) is depicted in figure 3.21. This circuit has certain disadvantages. One of these is the lack of common circuits for the 2 outputs, circuits which may lead to a more efficient design. Moreover, the two implemented functions are obtained by using completely different gates, as AND/OR respectively XOR, which also lead to too much area for such simple functions. Finally, an important problem may appear by using 3 or more input gates, having very large time delays, which is undesirable when output Carry is involved in the critical propagation path.

A more compact and faster variant could be acquired by starting from the truth table of the full adder (table 3.3) and writing the relations as follows:

$$S = a \oplus b \oplus C_i \tag{3.8}$$

and

$$C_o = ab + C_i(a \oplus b) \tag{3.9}$$

The expression for the Sum output is the same as that given by relation (3.7), but the Carry output is described in a different manner, so that we can use two half adders in order to generate the functions defined by the expressions (3.8) and (3.9), as it is showed in figure 3.22.

The logic expressions for the half adder outputs are defined as follows:

$$S = a\bar{b} + \bar{a}b = a \oplus b; C = a \cdot b \tag{3.10}$$

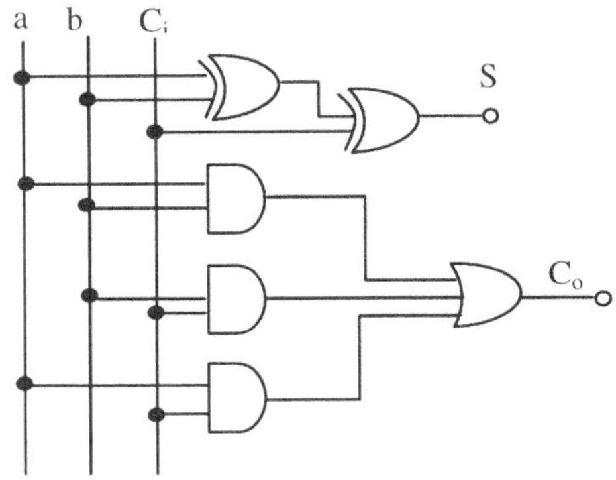

Fig. 3.21 The one bit full adder

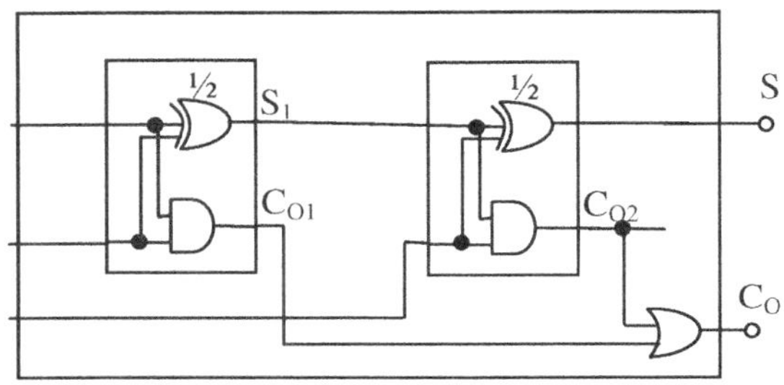

Fig. 3.22 One bit full adder using two half adders and one OR gate

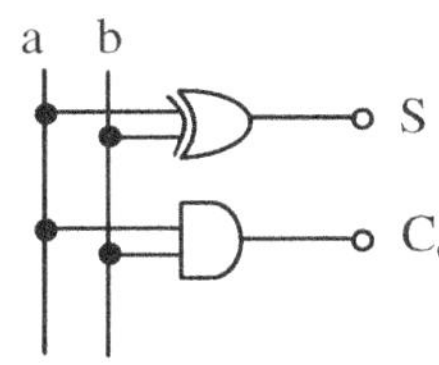

Fig. 3.23 The one bit half adder

Using relation (3.10), it results the direct implementation depicted in figure 3.23.

The signals propagation can be followed in figure 3.22. The first half adder provides the outputs:

$$S_1 = a \oplus b \;; C_1 = ab \tag{3.11}$$

The second half adder provides the following outputs:

$$S = S_2 = C_i \oplus S_1 = a \oplus b \oplus C_i \tag{3.12}$$

and

$$C_o = C_{o1} + C_{o2} = ab + C_i(a \oplus b) \tag{3.13}$$

Relations (3.12) and (3.13) show us that the circuit described in figure 3.22 implements the sum arithmetic operator, this configuration finding its use in the parallel processing ripple carry structure.

3.3.2.2 Parallel Processing Adder

The one bit adder described above is capable of adding numbers having 1 bit length and a supplementary bit named C_i (Carry input). When the input words' length is larger than 1, they have to design a structure composed of several adder cells of one bit each, parallel or serial connected.

The parallel processing adder for n-bit words consists of „n" primary cells, which are one bit cascaded full adders. Every stage receives the Carry input signal from the previous one and generates (to the next one) the Carry output for the next one, as it is shown in figure 3.24.

All the „n" stages receive simultaneously the input signals a_i and b_i, but the Carry signal is delayed in every stage, the total delay being as high as the stage rank is greater. Therefore, the final result is a consequence of the total number of stages. As the length of the added words is greater, the operation will need more time to be completed. A logic system detects the end of processing and allows the result to be loaded in a register.

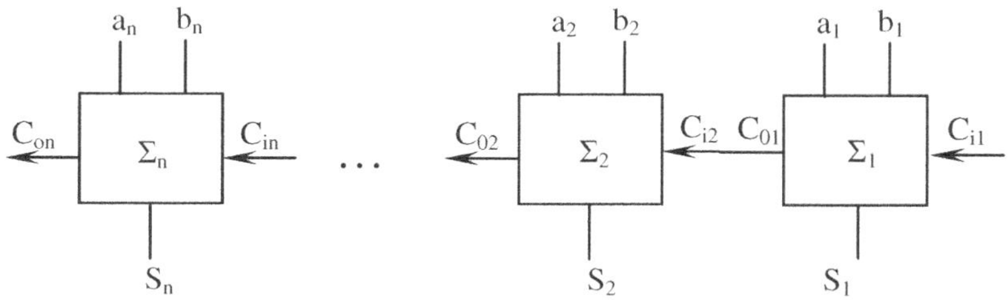

Fig. 3.24 The block scheme of the parallel adder

The delay time in the worst case (critical propagation path) is given by the above relation:

$$T_{adder} = (N-1)\, t_{carry} + t_{sum}$$

Where t_{carry} is the propagation time through one stage and t_{sum} is the delay time of the output signal Sum of the last added cell (Rabaey et al. 2003).

Two remarks are concluded:

1. The total delay time of a parallel adder is proportional with the length of the added words, which is a big disadvantage when dealing with words having great length.

2. The influence of the Carry propagation is stronger for the total delay time than that of the Sum signal and this influence increases also with the number of bits of summed binary words.

The above conclusions lead to the following design tasks:

1. We have to find architecture solutions for the adder cell taking into account the optimization of Carry signal more than that of Sum output.
2. Multiple bits parallel architectures that are able to obtain less dependence between the total delay time and the words length must be found.

3.3.2.3 CMOS Gates Full Adder

One solution is to directly implement the relations (3.6) and (3.7), which can be written as follows:

$$C_o = ab + bC_i + aC_i = ab + C_i(b + a) \tag{3.14}$$

Function S cannot be minimized, but it can be described by using the XOR operator:

$$S = a \oplus b \oplus C_i = a \cdot \overline{b} \cdot \overline{C_i} + \overline{a} \cdot b \cdot \overline{C_i} + \overline{a} \cdot \overline{b} \cdot C_i + a \cdot b \cdot C_i \tag{3.15}$$

This relation can also be written:

$$S = \overline{C_o} \cdot (a + b + C_i) + a \cdot b \cdot C_i \tag{3.16}$$

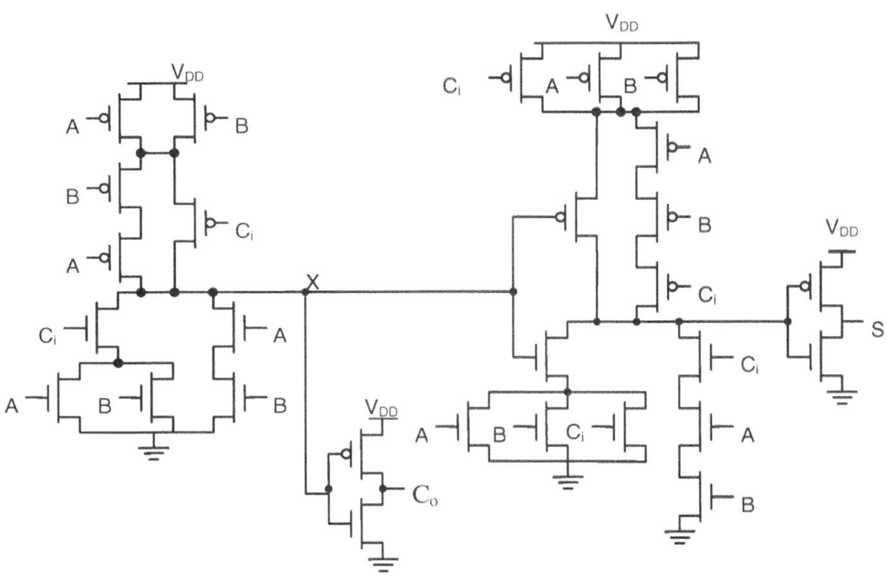

Fig. 3.25 One bit full adder made with CMOS logic gates

The circuit which directly implements the logic functions described above is given in figure 3.25.

We notice that the Carry Out signal is first generated and then it is used to acquire the signal Sum. The Carry delay time corresponds to that of 2 inverters propagation time, one of these inverting the signal X, and the second time is acquired by adequate sizing of the transistors composing the inverter having the output X. For instance, the N channel transistors are selected with the size 2/1 and P channel ones with 4/1, so that the circuit which obtains signal X has the delay time of a single inverter. The total number of transistors for this scheme is 28.

3.3.2.4 The Mirror Adder

The mirror adder is a circuit having improved performances, which implements the same logic functions described in equations (3.14) and (3.16), but in a different manner than the one given by the design described above (Rabaey et al. 2003).

So far the structure of figure 3.26 has two modules with transistors N and P, without a complementary structure, but the N channel transistors network is repeated identically, in mirror, in the P channel transistors network.

This circuit has some important advantages, one of them being the fact that in the circuit used to acquire the Carry signal all the paths have no more than 2 serial channels. Moreover, only the transistors situated on the Carry out path must be optimized as dimensions, while those placed on the Sum path could have minimum size. The total number of transistors used is 24.

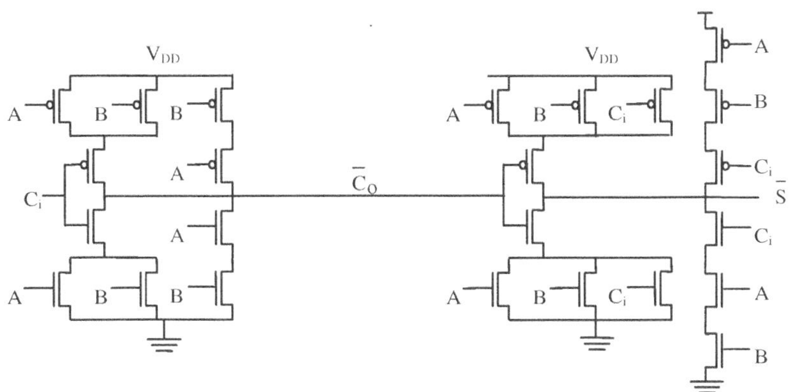

Fig. 3.26 The mirror adder

3.3.2.5 Full Adder with CMOS Transmission Gates

The employment of transmission gates in order to implement the one bit full adder is based on the following principles (see fig. 3.27).

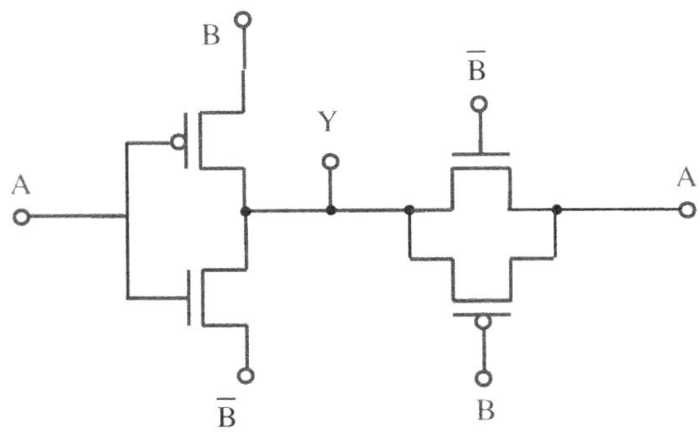

Fig. 3.27 XOR operator implementation with CMOS transmission gates

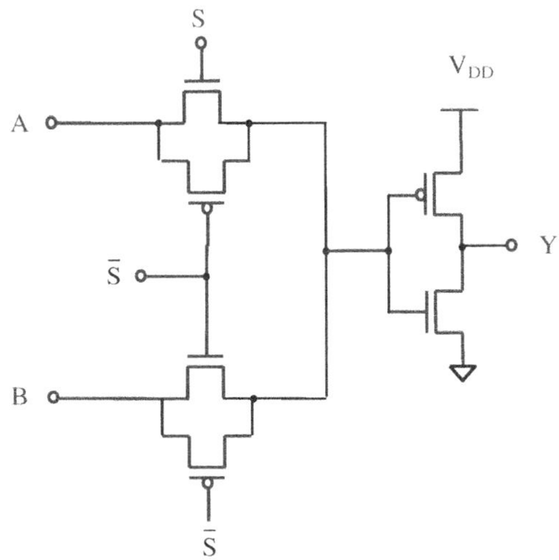

Fig. 3.28 Two inputs MUX implementation with CMOS transmission gates

When B is equal to "1", the inverter transmits the inverted A and the transmission gate is turned off, therefore Y = B $\bar{A}$. When B=0, the inverter is blocked and the transmission gate transmits A, therefore Y= $\bar{B}$ A. In other words Y= B $\bar{A}$ + $\bar{B}$ A = A $\oplus$ B. If we replace B with $\bar{B}$, we obtain the NOR operator.

Another application based on transmission gates is the 2 bit multiplexer, depicted in figure 3.28. In this case, if S=1, the signal $\bar{A}$ is transmitted at the output, while if S=0, the signal $\bar{B}$ is transmitted at the output. The result is Y = S $\bar{A} + \bar{S}\ \bar{B}$.

The full adder depicted in figure 3.29 is made using the principles described above and starting from the relations defining the Propagate (P) and Generate (G) terms:

$$P= a \oplus b \ ; G = ab$$

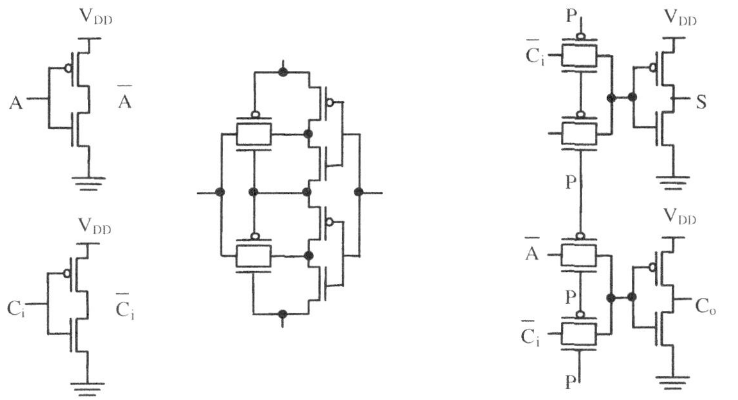

Fig. 3.29 One bit full adder with CMOS transmission gates

Carry and Sum signals are given by the relations:

$$S = P\oplus Ci = a \oplus b \oplus Ci \tag{3.17}$$

and

$$Co = G + P \cdot Ci = ab + (a \oplus b\,)Ci \tag{3.18}$$

The main advantage of this transmission gates implementation is that all the paths have small resistance, each of them consisting of no more than two serial channels. Also, certain circuits are used in common, including the XOR operators (Rabaey et al. 2003).

3.3.2.6 Serial Processing Adder

The serial processing adder uses a single 1 bit full adder as it is shown in figure 3.30. The sum is processed bit by bit. The n-bit words at the input are loaded into the serial registers A and B. The result is shifted to the Sum register S.

On every clock edge (Ck), one bit of the same rank of every input word is summed with the adding of the carry resulted from the last operation. In the same time the resulted bit is shifted in the Sum register. As a result, after n clock pulses we have the n bits Sum word in the Sum register.

Compared to the parallel processing system, the serial one is more profitable because only one 1 bit full adder is used, but the processing speed is smaller because we need "n" clock cycles in order to add "n" bits words.

3.3.2.7 Conventional Binary Multipliers

As it is well known, multiplication can be made by using repeated additions. The pen and paper multiplication method can be resumed as follows. We consider each bit of the first term. If this bit is "1", the second term is added and then shifted; if not, a "0" is added and then it is shifted.

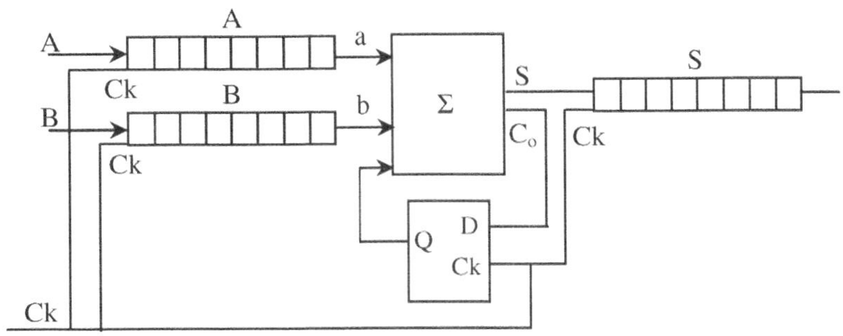

Fig. 3.30 Serial adder block diagram

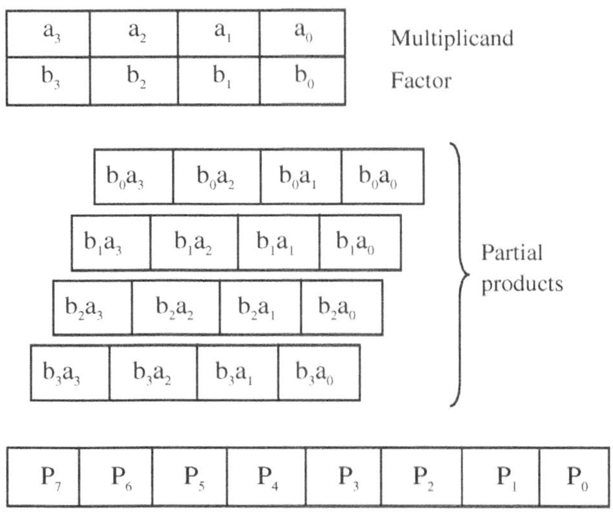

Fig. 3.31 Basic principle for the combinational multiplier

In multiplying by hand the partial products are shifted to the left and finally an addition is made. The same result could be obtained by adding after looking for each bit and then shifting the sum to the right.

We need registers to keep the information of the multiplicand, the factor, the sign and the product. It must be noticed that if both operands have "n" bits, the result needs a "2n" bits register.

The end of calculus is indicated by a counter which registers the number of bits of the factor. For example, the multiplicand (1011) is memorized in register A, the initial sum (0000) is kept in register B, while the factor (1001) is memorized in register C.

The multiplication can be made by using a combinational circuit which works as it is shown in a truth table which describes the product of “2n” bits as a logic function of inputs A and B, both of them having a length of “n” bits.

Figure 3.31 shows the working principle for a combinational multiplier for 4 bits words. The operands are $A=(a_3a_2a_1a_0)$ and $B=(b_3b_2b_1b_0)$. The partial products are composed of the results provided by the AND gates having the entries a_i and b_j. The result has 8 bits and it is obtained by adding all the shifted partial products.

The practical combinational multiplier is depicted in figure 3.32. Here we can see that the circuit consists of only 16 AND gates and 12 one-bit full adders. The products a_i , b_j are noted with P_{ij}. The final result is the binary word $S=(S_7S_6S_5S_4S_3S_2S_1S_0)$.

In the general case of “n” bits words, the partial products can be computed in a single time unit using a number of “n^2” AND gates. The partial products are added using the one-bit full adder. The multiplier speed depends on the delay time via the logic gates and adders.

The worst case (the slowest speed) will be obtained on the critical path which contains the circuits placed along the right diagonal and the lowest horizontal line. This critical path contains a number of 20 one-bit adders. If each adder has a delay time of t_p, it results a total value of 20 t_p (Rabaey et al. 2003).

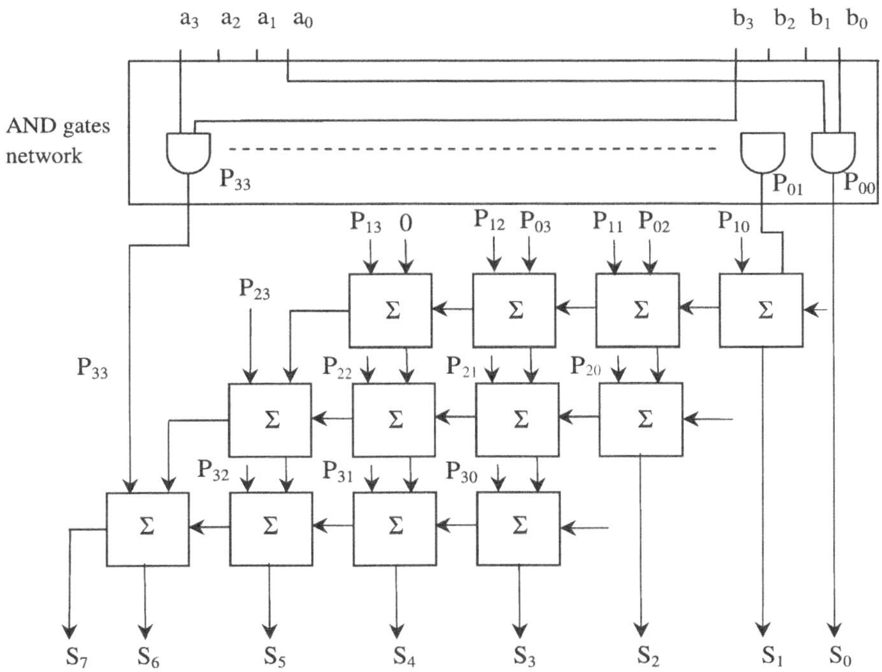

Fig. 3.32 Binary multiplication with combinational logic circuit

The multiplier working speed can be improved as it is showed in figure 3.33, the carry output of "n" adder not being applied to the next adder placed in the same row, but to another one, positioned in the upper next row. The critical path contains now only 14 adders, which is an important improvement, respectively decreasing with 30% the propagation time.

In the general case of having two numbers of "n" bits multiplied, the delay on the critical path is given by a number equal to (n-1) + (n-1)=2(n-1).

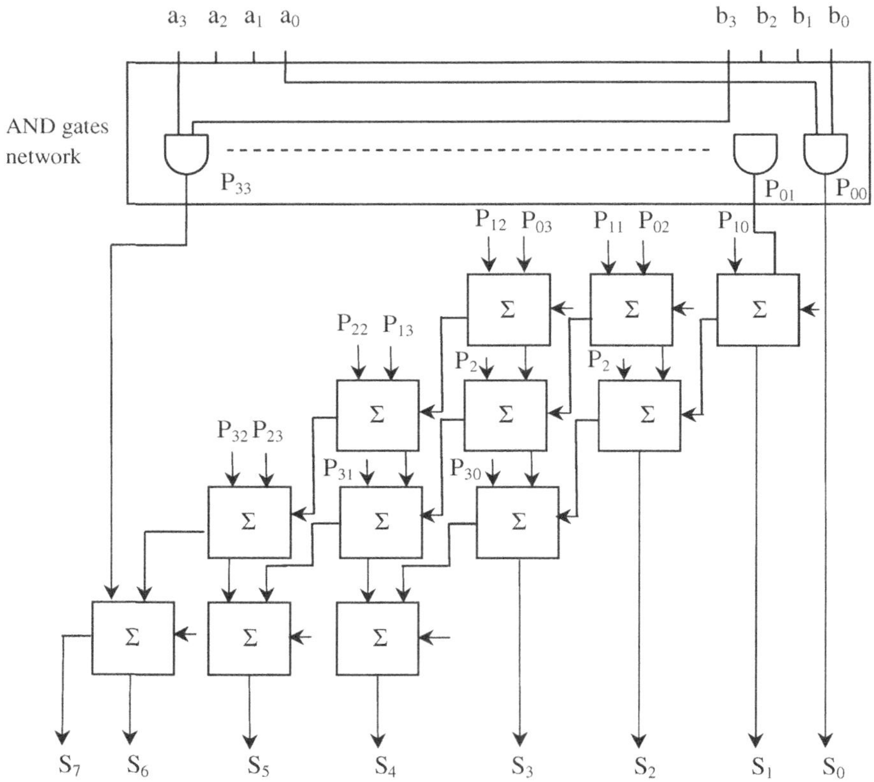

Fig. 3.33 Acceleration binary multiplier

3.3.3 Arithmetic Circuits Designed with Evolutionary Algorithms

3.3.3.1 Full Adders Design

There are various methods proposed during the last decade regarding the evolution arithmetic circuits, adders or multipliers (Miller et al. 1997; Vassilev et al. 2000; Coello et al. 1997; Aoki et al. 1999).

One of the first approaches was made by (Miller et al. 1997) in order to discover new designs for arithmetic circuits using evolved methods.

In this manner, they obtained the circuit showed in figure 3.34, where the output Sum is the same as in the conventional adders, but the carry is provided with the help of two multiplexers. These circuits are widely encountered in FPGA Xilinx structures.

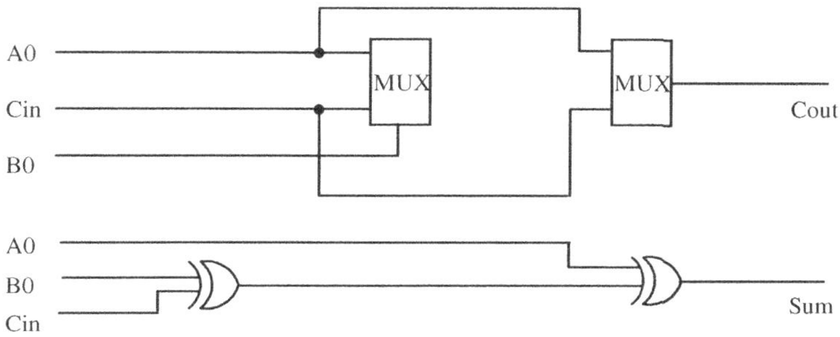

Fig. 3.34 One-bit full adder with carry

Another structure was obtained when the geometry was constrained to 2 rows and 2 columns and the GA runs twenty times (over 2000 generations), this being an optimal solution (see fig. 3.35).

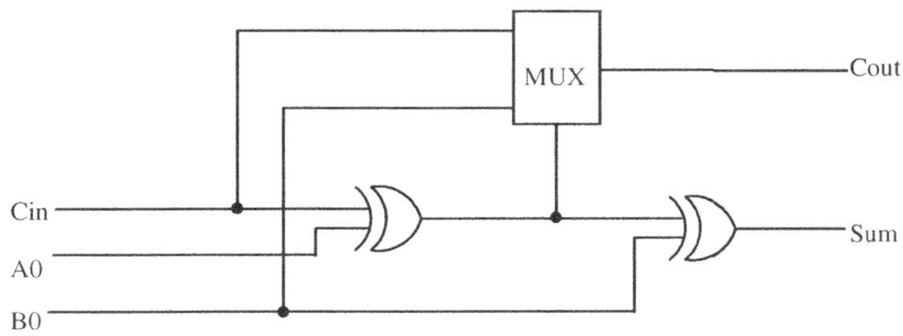

Fig. 3.35 One-bit full adder with carry – the evolved optimal solution

The next step was to extend the adder's capacity to more than 1 bit. The specific of this design is that the 2 bit adder is viewed as a black box having 5 inputs and 3 outputs (fig. 3.36). This approach is different from the parallel processing earlier described scheme, in which we have the same cell multiplied by "n" times.

Two approaches are presented: the first one having the sum and the carry signals separated and the second one having these two outputs combined. The first circuit is depicted in figure 3.37.

Fig. 3.36 Two-bit full adder with carry

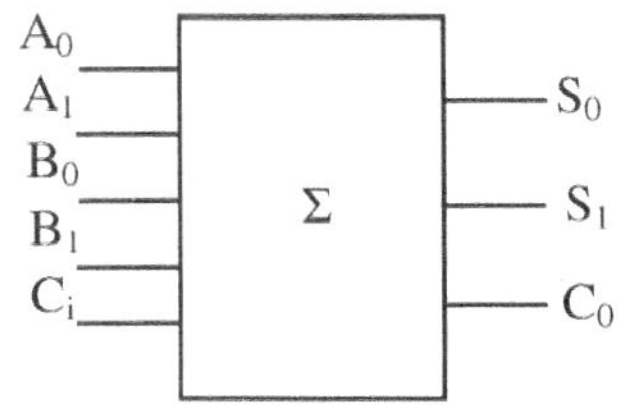

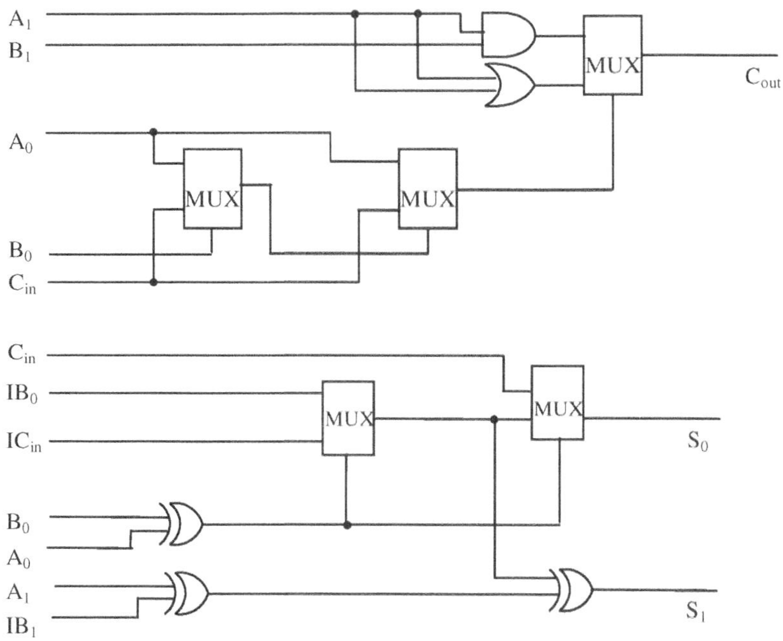

Fig. 3.37 Two-bit full adder with carry – the evolved solution with separated components

The second implementation brings into light the evolutionary methods' advantage, in this case the structure obtained being more compact as it is showed in figure 3.38.

This scheme is very similar with the ripple-carry adder and the generalization could be obtained by directly connecting the one-bit adders showed in figure 3.24.

3.3.3.2 Gate-Level Evolutionary Design

One contribution uses the Evolutionary Algorithm inspired by the Cartesian genetic programming (Gajda and Sekanina 2007). This method is proposed to be applied for arithmetic circuits design, but also for majority of the other type of circuits.

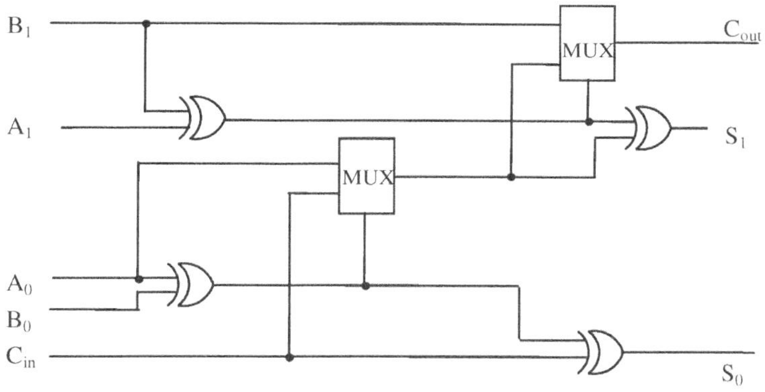

Fig. 3.38 Two-bit full adder with carry – the evolved solution with combined components

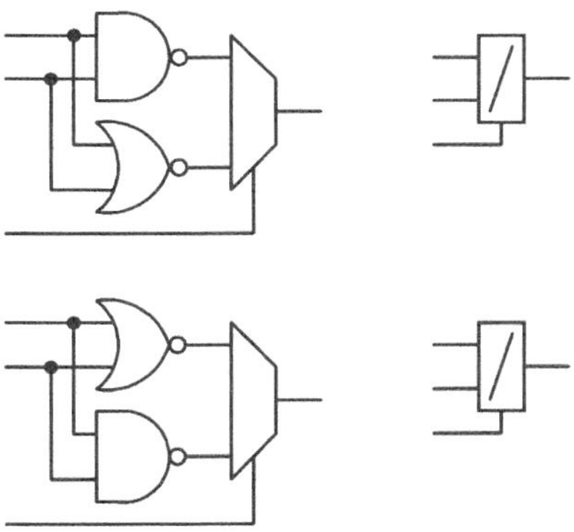

Fig. 3.39 Evolved 1-bit full adders

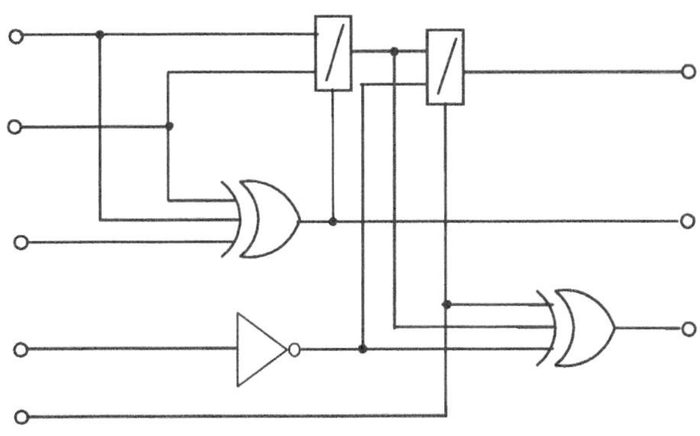

Fig. 3.40 Evolved 2-bit adder

This approach is based on a cell built with NOR/NAND gates and described in figure 3.39 (Gajda and Sekanina 2007). The basic idea is to use other structures than the fundamental cells (NAND or NOR gates) in order to obtain better results.

In the case of the evolved one bit full adder, different solutions are obtained, all of them having 22 transistors for all combination of parameters. This result is better than the mirror adder with its 24 transistors.

Also some results are presented for the 2-bit adder. The evolved solution shown in fig. 3.40 uses two NOR/AND gates, two 3-inputs XOR gates and an inverter, these leading to a number of 42 transistors, less than the 48 transistors needed in the case of cascading two 1-bit full adder in a ripple carry scheme.

The evolved solution was optimized from the point of view of number of transistors and the delay at the gate level. But some parts need to be discussed in what concerns the delay optimization at transistor level, power consumption, placement and routing aspects. This approach is also remarkable for the very short time needed to design this sort of circuits (154 seconds for the 2-bit adder in a single run of EA which uses 60 programmable nodes and produces 1,000,000 generations) (Gajda and Sekanina 2007).

3.3.3.3 Binary Multipliers Designed with Evolutionary Algorithms

Some new evolutionary methods of evolving digital multipliers were reported (Miller et al. 1997; Vassilev et al. 2000, Coello et al. 1997). (Vassilev et al. 2000) proposed a new approach applied to the three-bit and four-bit multipliers.

The digital circuit is encoded using a genotype-phenotype mapping defined with the help of a cells array. Each cell is a two-input logic gate. The logic gates allowed are AND, AND with one input inverted and XOR.

The genotype is a linear string of integers and it is defined by four parameters of the array: the number of allowed logic functions, the number of rows, the number of columns, and the levels-back. While the first parameter defines the functionality of the logic cells, the next three parameters determine the layout and routing of the array. The number of inputs and outputs of the array are specified by the objective function (Vassilev et al. 2000).

The method is used to design some several binary multiplier circuits such as the three and four-bit multipliers.

One example designed using this method is described in figure 3.41. It is a three x two multiplier made using only 9 AND gates, 4 XOR gates and 1 inverter. The total number of 2 input gates is 13. Compared to the conventional design which requires 17 two input gates, the evolutionary method leads to a 23.5 % more efficient design.

The authors offer different other examples increasing the operands length from two to four. As the multiplier capacity increases, the results are also better than the conventional design, but the improvement becomes smaller when the bits number is increasing. In the case of a four x four multiplier, the resulted evolved circuit consists of 57 two - input gates and it is about 11 % more efficient than the conventional solution, which uses 64 two - inputs gates.

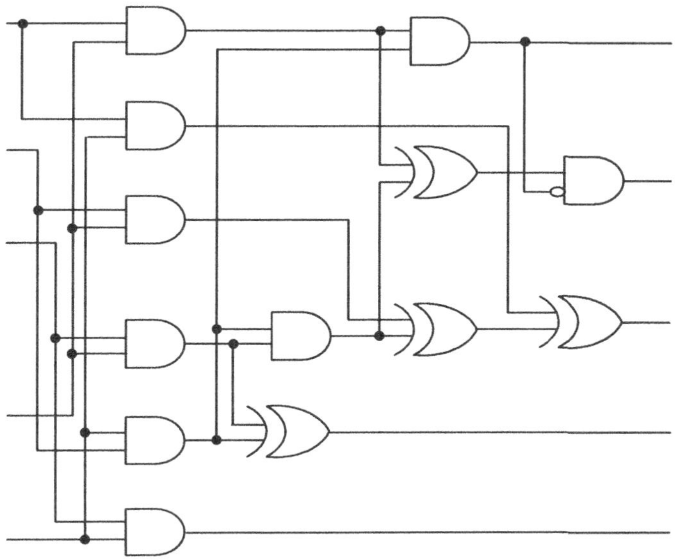

Fig. 3.41 The three-two-bit multiplier circuit

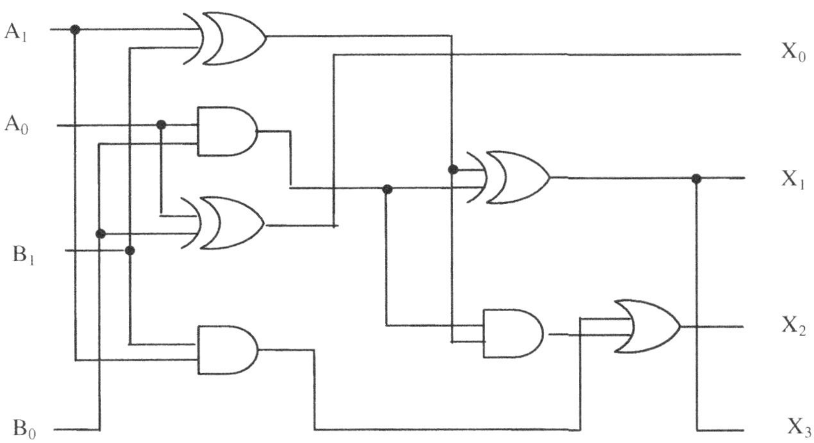

Fig. 3.42 The two-bit adder circuit

(Coello et al. 1996) also introduced a method based on GA in order to design combinational logic circuits, particularly arithmetic circuits. The solution for a 2-bit adder described in figure 3.42 contains only 7 gates, which is an important gain compared to the 12 gates used by human design.

Also, the authors evolved design schemes for 2-bit multiplier. The result is a circuit with only 7 gates, replacing the 16 gates given by the conventional design techniques. This result is showed in figure 3.43.

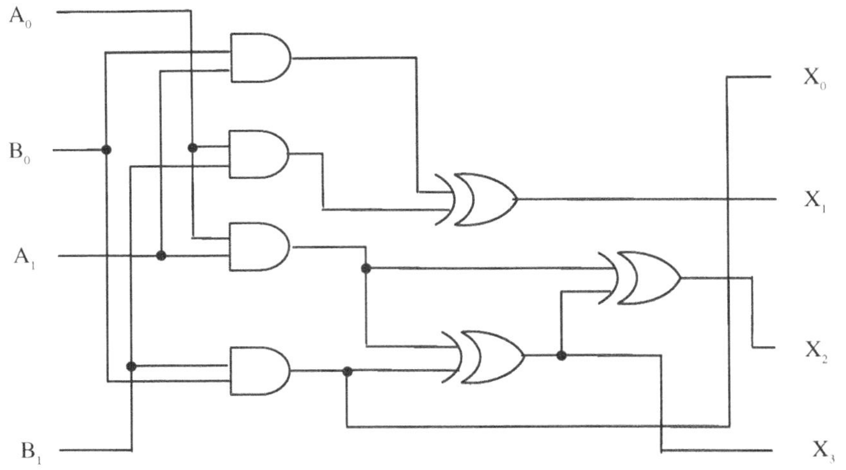

Fig. 3.43 The two-bit multiplier circuit

3.3.4 Concluding Remarks on Digital Circuits Evolutionary Design

First at all we have to mention that circuits designed by evolutionary methods are different from the conventional designs. That means that the same logic operation is obtained in a different manner and it is quite impossible to find such a solution by classical approaches.

One major difficulty appears when we try to evolve larger systems because the size of the truth table grows exponentially with the number of inputs (Miller et al. 1997). It results that it could be easy to evolve quite simple circuits.

Some of he above given examples show us that a small increase of circuit complexity, from a 3-bits to a 4-bits multiplier could lead to an important increase in complexity and much more difficulties.

When the truth table becomes too large, the evaluation of the fitness of chromosomes could lead to very slow evolution process. In case of the combinational circuits' evolution the evaluation time of a certain circuit grows exponentially with the increasing number of inputs. This is the so-named *Scalability of evolution*, which refers to the fact that evaluation time is the principal bottleneck of the evolutionary method (Sekanina 2006).

One explored solution was to evolve efficient designs for the small building blocks taking into account that the arithmetic circuits could be defined as modular structures. For example, we can evolve 4-bits structures and multiply the obtained structures to obtain a longer length bits adders or multipliers. The ripple-carry adder is a typical application. And this kind of solution is known from the carry look ahead blocks of 4-bits length used to design a much larger ripple-carry adder.

Torresen introduced a *divide – and – conquer* method in order to reduce the effects of S*calability of evolution* and evolve more complex digital circuits. This approach is also named *increased complexity evolution* and consists of a division of a system in smaller blocks (Torresen 1998). The evolution is first applied individually to these units and then the evolved blocks will be used in the evolution of a more complex system.

Torresen proposed this method to evolve prosthetic hand controller and the reported results were very good (Torresen 2002).

The incremental evolution was used by dividing a complex circuit into simpler components in order to evolve each of the small circuits and then rebuilding a new evolved complex system (Kalganova 2000). This approach was applied to different digital circuits, including 6 bits multipliers.

One interesting aspect is that evolution could open new possibilities in combinational logic design. It is well known that traditional logic synthesis techniques such as Karnaugh maps and the Quine McCluskey procedure generate sum-of-products solutions.

There are VLSI design techniques within the XOR gates and multiplexers are more preferred approaches and in this case the evolutionary method can offer better automated solutions.

(Sekanina 2006) made a synthesis about the evolutionary digital circuits design and about the complexity and innovation that can be obtained by using these unconventional techniques. The main approaches proposed to be used in digital design are: Gate-Level Evolution, Functional Level Evolution and Transistor Level. The main limitation of these techniques is the restriction of about 1000 bits for the chromosomes length. At this value, the needed computational power becomes too uncomfortable.

In the case of the Gate Level Evolution, the complexity of designed circuits is up to 100 gates (Sekanina 2006). In this case, all the inputs-outputs combinations are simulated and tested. In comparison, the transistor-level approach could be used only for small digital circuits. Only the functional-level approaches could lead to more complex and innovative solutions. In this case, because of the circuits' complexity, only a subset of all combinations is considered to be evaluated in the fitness function.

3.4 Reconfigurable Analogue Circuits in Mobile Communications Systems

3.4.1 Multi-standard Terminals for Mobile Telecommunications

The last decades highlighted an explosive growth of mobile telecommunications, including the most demanding requirements. Over the years the goal was to achieve higher data rates and to introduce new services (Silva et al, 2005).

The tendency is to support several standards in the same handheld device, with the power consumption and area restrictions (Silva et al, 2005).

The large variety of Radio Access Technologies (RAT) that exists worldwide is the result of the mobile communication systems' widespread in the last three decades. Two important families of standards are prevalent today on a global basis: the GSM system with its many variations and the CDMA-based family of standards, UMTS. Each of these systems has derivations that altogether with personal radio systems (Bluetooth), positioning systems (GPS), local and wide area data networks (VLAN, WiMAX) and military systems (TETRA) create an unprecedented diverse spectrum distribution of existing mobile communications standards. Carrier frequencies for these standards range from 400MHz to around 5GHz, while each standard has particular specifications in terms of modulation, analogue and digital signal processing and hardware requirements. These aspects make it more and more difficult to integrate all these radio interfaces into a single mobile terminal that can also offer sufficiently capable battery-powered hardware in miniature sizes (Rus and Hintea 2007).

Mobile communications systems have developed at an unprecedented pace in the last years, having a major influence in almost all aspects of modern citizens' every-day life. Worldwide the tendency is to widen the coverage area of mobile communications systems and to ensure interoperability in order for the users to benefit of global mobility. Major studies indicate that new services like global mobility may be offered under the premise of designing mobile communications systems reconfigurable at hardware level (Hintea et al, 2007).

In order to offer mobile communications user services adapted to their needs/requests, mobile networks must possess a high degree of intelligence. The intelligence is implemented at the application level through flexible software algorithms, and at physical level through a reconfigurable hardware platform whose parameters may change under software control. User terminals are the main impediment in the development of a completely reconfigurable mobile network due to the hard-to-meet specifications concerning battery life, cost, physical dimension (from the user's point of view) and power consumption, linearity and circuit integrability (from the designer's point of view) (Hintea et al, 2007).

Many new wireless standards were introduced during the last decades. The European Global System for Mobile Communications (GSM) and its copies DCS and PCS, dominate standards for cellular communications. In general the 2G wireless cellular systems are used depending on the geographic region. There are some other important 2G standards like PDC (Personal Digital Cellphone) or IS-95 (American North American Digital cellular) (Stehr et al, 2003).

The 3G wireless system is known as UMTS (Universal Mobile Telecommunications System) and it is able to handle data rates of up to 3.84 Mb/s. This performance is obtained by using wide-band width signals with Code-Division for Multiple Access (W – CDMA or WCDMA). The bandwidth for UTMS – signals is approximately 4 MHz and it will be increased up to 15 MHz.

In table 3.4 are listed the dominated 2G and 3G systems (Mak et al, 2007).

At the same time, the services requested by mobile users, may they be civilians or militaries, are new and diverse in the last years; most common commercial services requested are wireless high-speed web browsing, multimedia streaming, position localization; on the military side, the keyword is integration, as the large

Table 3.4 The dominated 2G and 3G systems

	GSM	**DCS**	**PCS**	**WCDMA**
Modulation	GMSK	GMSK	GMSK	GMSK
Frequency Band	890-960 MHz	1710-1850 MHz	1880-1930 MHz	1920-2170 MHz
Channel band-width	200 kHz	200 kHz	200 kHz	3.84 MHz
Bit rate	270 kb/s	270 kb/s	270 kb/s	3.84 Mb/s

number of separate and incompatible radio interfaces impede seamless communication between the various military actors. Common to the civilians and militaries is not only that the services they need must be accessible to users at any time, but most importantly they must be approachable anywhere and in any mobile environment through one intelligent wireless device.

The Software Defined Radio (SDR) technology offers, in this context, the framework for the study of future mobile communications systems that will offer seamless mobility and diversified services, accessible worldwide with one mobile terminal (Mittola J, 1995; Tuttlebee W, 2002). An SDR-enabled radio is envisioned, in its widest acceptance, to perform the entire signal processing in a general-purpose processor, capable of implementing the specifications of any present and future mobile standard. However, since the SDR technology was first coined by the US Department of Defence in the late 1980's, insignificant progress have been made in developing a practical implementation of an SDR-compatible transceiver, mainly due to technology limitations. Extensive studies exist at the application/service level, but possible practical hardware implementations of software-radios have been studied much less; one example in this direction is a state-of-the-art implementation of a multifunction antenna suitable for SDR military applications (Perciante A, 2006).

Due to insufficient analogue-to-digital converter capabilities with state-of-the-art ICs, sampling waveforms at GHz carrier frequencies is not possible. For this reason analogue circuits are the bottleneck in the development of a practically-feasible SDR transceiver. Few theoretical studies have been published on reconfigurable analogue circuits (Maurer L et al. 2005) and no generalized methods in developing such circuits exist. (Rus C et al. 2007) presented a digitally programmable and reconfigurable analogue array for the intermediate frequency (IF) filter stage of a wireless transceiver. The circuit is capable to implement analogue filters of various orders, approximations and methods of synthesis.

The goal is to allow an easy interoperability between wireless standards and the transition from 2G and 3G generation mobile equipments, and also to provide new services. The principal problem is to support several standards in the same hand-held device.

A Multi-standard terminal must be capable to operate with a multitude of RATs (Radio Access Technologies) at the same time. This could be possible using a structure including RF transceivers with concurrent paths. In this case they can design as many channels as necessary, one for each standard. But this solution has very high costs because of using a lot of materials and also too much area occupied for electronics.

3.4.2 Reconfigurable Multi-Standard Analogue Baseband Front-End Circuits in Mobile Communications Systems

In order to offer to mobile communications user services adapted to their needs/requests, mobile networks must possess a high degree of intelligence. This intelligence is implemented at application level through flexible software algorithms, and at physical level through a reconfigurable hardware platform whose parameters may change under software control. User terminals are the main impediment in the development of a completely reconfigurable mobile network due to the hard-to-meet specifications concerning battery life, cost, physical dimension (from the user's point of view) and power consumption, linearity and circuit integrability (from the designer's point of view) (Agnelli F et al 2006).

The **S**oftware **D**efined **R**adio technology offers the context for the research of reconfigurable mobile systems. The **S**oftware **R**adio concept refers, in its largest understanding, to unrestricted programmability and re-configurability of equipment, functions and services in a mobile communications system, from the application layer down to the hardware platform. Unlike today's communications systems, a software radio transceiver not only sends/receives radio signals but also selects the multiple access technique, the optimum available frequencies, modulates multiple signals corresponding to different standards and dynamically cancels the interferences (Maurer L et al. 2005). A reconfigurable mobile communication system uses a completely reconfigurable mobile transceiver that is completely reconfigurable in which the A/D and D/A converters placed right after the antenna allow all radio functions to be implemented in software over a general purpose processor (fig. 3.44) (Baschirotto A et al 2006).

Sampling radiofrequency signals at their carrier frequency according to the ideal model eliminates the bulky analogue circuits that are not amenable to reconfiguration and with high power consumption. Despite the many advantages offered by the ideal architecture (no analogue circuits, high flexibility and low cost), the technological limitations related to the power consumption of a practical A/D

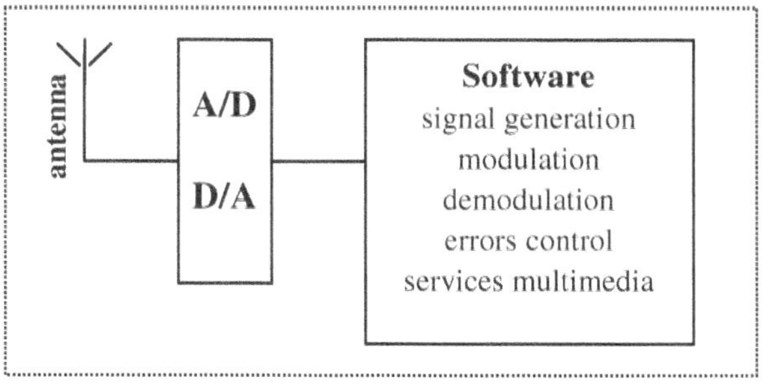

Fig. 3.44 Principle diagram of a software radio transceiver

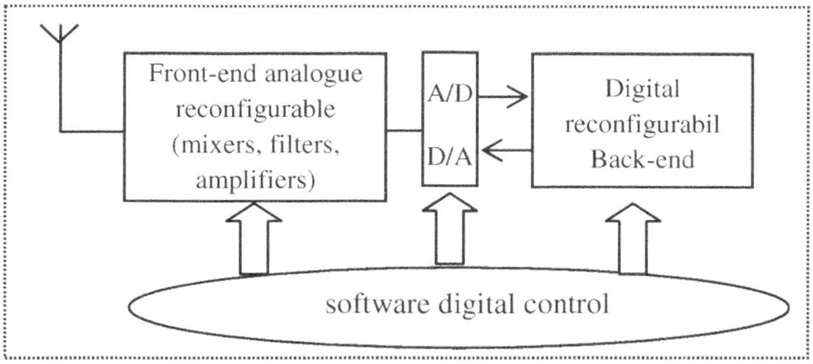

Fig. 3.45 Basic structure of a practical software defined radio transceiver

converter capable of sampling carrier signals (900 MHz, 1,8 GHz, 2,4 GHz) impose an analogue front-end that must adapt the received signal to the specifications of the A/D converter through multiple frequency conversions, amplification stages and filtering (fig. 3.45) (Maurer L et al. 2005).

The research for the development of a reconfigurable mobile terminal is justified by the need to make one compatible device with all the data and voice mobile standards in order to offer the user unrestricted mobility and diversified services. Even the utmost modern transceivers are significantly different from the ideal model depicted in fig. 3.44. An existing mobile terminal offers compatibility with a limited number of standards, usually three, and the mode of operation (in one of the three modes) is selected by switching between similar circuits but tuned for different operating parameters (fig. 3.46).This aspect implies a high hardware redundancy by deactivating a circuits chain (filters-mixers-amplifiers) between the antenna and the analog-to-digital converter, while the similar chain tuned for slightly different parameters is activated (Strohmenger K et al. 2003).

As the user needs for mobility grow and the required services are more and more diversified, making a terminal compatible with more standards cannot occur by inserting supplementary circuits chains in the transceiver, similar to the already existing ones. The high cost of such a mobile terminal, the high hardware complexity and most important, the limited compatibility (only with those standards that are implemented in dedicated hardware) are decisive arguments against such an approach. The solution lies in using only one chain of circuits that is able to adapt by topological reconfiguration and parameter programming to the specifications of the service or standard solicited by the user (fig. 3.46) (Baschirotto A et al 2006).

Due to the fact that – according to the current technological development – a transceiver is compatible with a limited number of standards (dual or tri-band terminal), the research focusing on reconfigurable circuits is the main study direction for the development of any-standard mobile terminal.

The design of analogue reconfigurable and programmable circuits that allow the implementation of a flexible transceiver front-end is presented in this chapter.

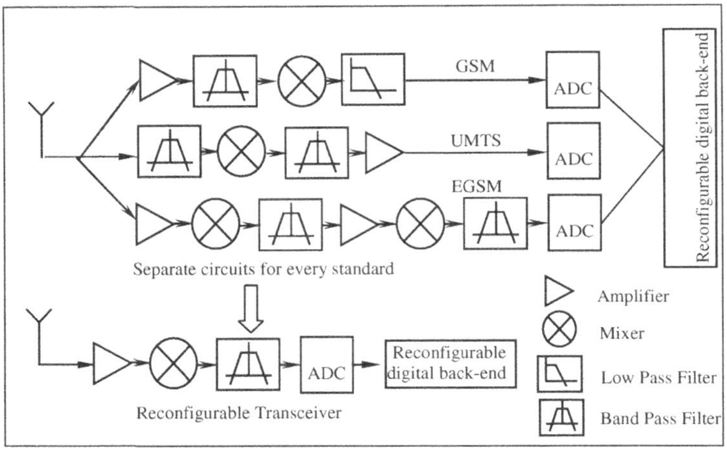

Fig. 3.46 Comparison of hardware resources between a current generation mobile transceiver and a reconfigurable one

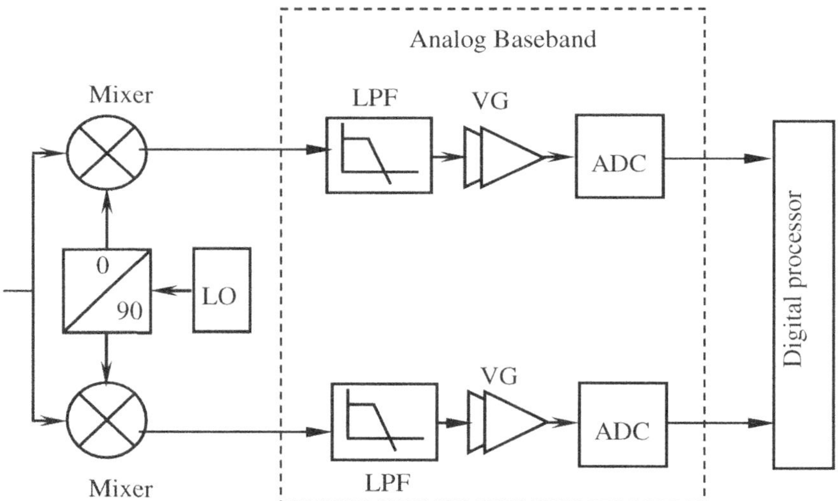

Fig. 3.47 Block diagram of analogue baseband front-end circuits

One of the fundamental conditions for implementing a reconfigurable transceiver is the reduction of number of analogue front-end circuits (Maurer L et al. 2005).

From this point of view, our objective is the study of the practical methods of implementing the three fundamental blocks from a mobile transceiver: the high frequency amplifier, the band-pass filter and the mixer. Nowadays the trend is to integrate all circuits on a single chip.

From this perspective, under the trend of evolution towards wireless, making one mobile terminal compatible with all mobile communications standards is crucial for the evolution of wireless communications systems (Baschirotto A et al 2006).

If we want to use the same terminal to receive many standards, the RF front-end must have reconfigurable building blocks. This is possible in the case of direct – conversion receiver (DCR), where are only low-pass filters in the analogue baseband and is suitable to be applied to multi mode receiver (fig. 3.47).

The low-pass filters have a good ability to be reconfigured from the point of view of bandwidth, gain and linearity (Stehr U et al. 2003).

3.4.3 Reconfigurable RF Receiver Architectures

3.4.3.1 Superheterodyne Receiver

The classical architecture used for receiver architectures is the heterodyne structure depicted in fig 3.48. The signal band is first translated down to an intermediate frequency (IF) which is usually much lower than the frequency of the RF signal. The requirements of the channel select filter (BPF) are not so restrictive.

The mixer is followed by an image reject filter. The mixer creates two bands located above and below the local oscillator frequency. The consequence is that we need a filter to reject the image frequency just after the first mixer. If frequency IF is large, the image filter characteristics will be more relaxed.

The disadvantages are the need for a large number of external components and also the complexity of the structure. We need external components because the channel selection filter is designed for a high IF value and most of the filters are made today with Surface Acoustic Wave (SAW). Another consequence is that this solution is not good from the point of reconfigurability (Maurer et al. 2005).

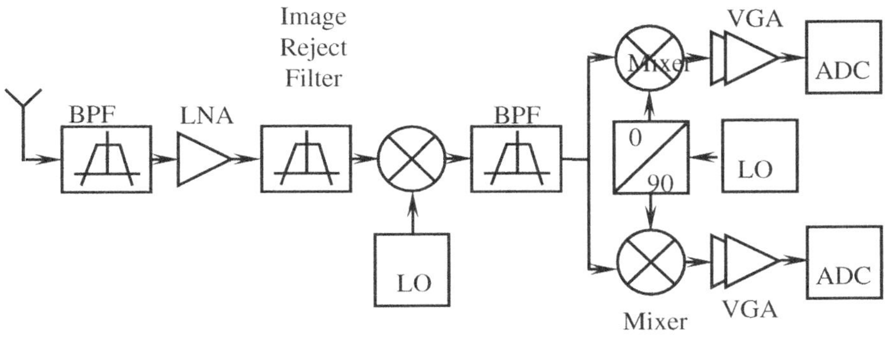

Fig. 3.48 Heterodyne receiver structure

This architecture dominated for decades the RF receiver design, but because of the drawbacks like high power consumption, external components, high costs, low level of integration in CMOS technology, low possibility of configurability, other solutions became more suitable.

3.4.3.2 Direct - Conversion Architecture

The homodyne structure, also called **zero – IF** or **direct – conversion architecture**, is obtained from the heterodyne architecture when we consider an IF frequency equal to zero (fig. 3.49). The LO frequency is equal to carrier frequency and thus, IF is equal to zero.

There are two principal advantages over a heterodyne structure. First, there is no need of an image filter because IF frequency is 0. Second, the IF filter and the next stages are replaced with low-pass filters and baseband amplifiers, which can be easily integrated. Then the base band filters and ADC could be programmable and realized in the CMOS technology.

All the filters could be integrated and there is no need for an external filter, as in the case of heterodyne receiver. If we need a multimode or multi-band application, it is possible to change the bandwidth of integrated low-pass filters.

The disadvantages are important too: first, because of the limited isolation of the local oscillator to the RF path there is an important DC – offset. Another problem could be the amplitude and phase mismatches between I and Q paths. Also, this structure leads to the apparition of a flicker noise during the signal processing at base-band. The base band filter linearity and noise are more demanding than for heterodyne receiver.

The direct-conversion receiver became the predominant architecture for wireless applications (Maurer et al. 2005)

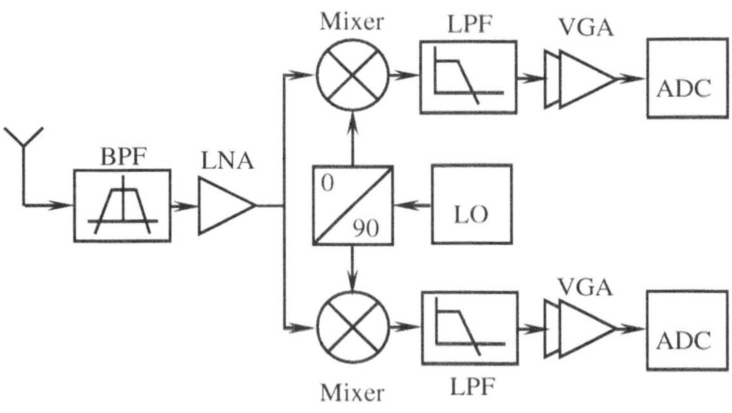

Fig. 3.49 Homodyne receiver structure

3.4.3.3 Low IF Architecture

In the Low – IF topology the RF signal is translated to an intermediate frequency closer to the baseband frequency (Silva et al. 2007). The advantage consists in keeping the same level of integration as the direct-conversion. The band-pass filters are more difficult to be designed and the ability for multi-standard is affected (Silva et al. 2007).

This receiver keeps all benefits for a lower-power and integration solution in CMOS technology, avoiding the principal problems of zero-IF architecture: dc-offsets, flicker noise and second harmonic distortion (Maurer et al. 2005).

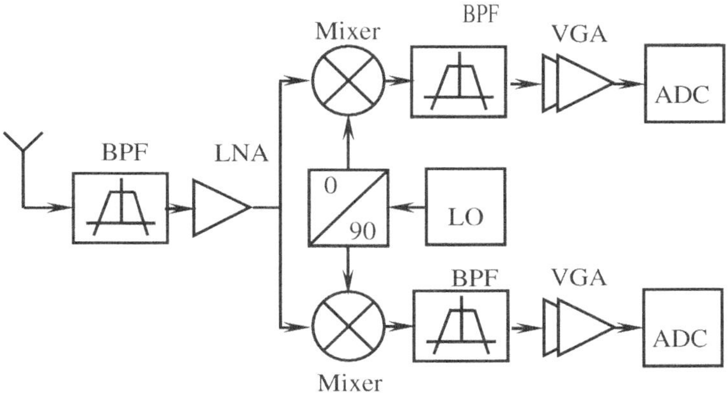

Fig. 3.50 Low IF front-End Architecture

In this case the IF bandwidth is selected using a polyphase filter stage. The next VGA (Variable Gain Amplifier) improves the amplitude of the signal to the required level for the ADC. All these stages (BPF, VGA and ADC) are suitable for reconfigurable design (fig. 3.50).

The low – IF architecture works well for narrowband baseband signals. For the case of wide band signals a better solution is the direct-conversion.

3.4.3.4 Software Defined Radio

The Software Defined Radio (SDR) concept was initiated by J. Mitola in the early '90s (Mitola 1995). The ideal SDR is depicted in fig.3.51. The receiver consists of a low-noise amplifier (LNAS), followed by an anti-aliasing filter and by the analogue-to-digital converter (ADC). That means the received RF signal is converted and processed in the digital domain with the help of a Digital Front End (DFE).

The basic idea of this architecture was to move as much signal processing as possible from the analogue domain to the digital one, where all the operations could be made in software on a digital signal processor (DSP).

But unfortunately this architecture is not feasible for the current mobile systems, because the required performance for the RF digital processing is too demanding. For example, the power consumption is excessive when you have to sample at RF frequency and the dynamic range required for the ADC is not feasible today (Dellsperger et al. 2006).

Shortly speaking, Software Defined Radio (SDR) is a reconfigurable transceiver. The receiver could be programmable by software, defining the radio parameters and which can be upgraded to face the requirements of new future protocols.

The idea of SDR is to convert the radio signal from analogue to digital as near as possible to the antenna, allowing the use of Digital Signal processing DSP or other digital programmable hardware.

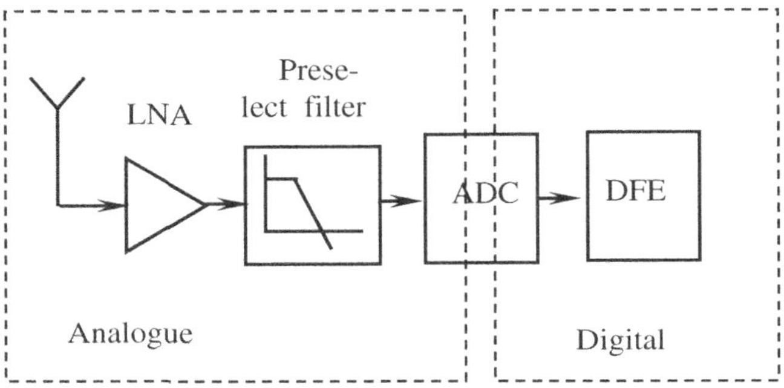

Fig. 3.51 Ideal SDR receiver

Even this architecture is not practical now, the basic idea is used in many cases: to shift as much signal processing as possible to the digital domain. In this way another advantage will be touched: the digital blocks can be made configurable more easily.

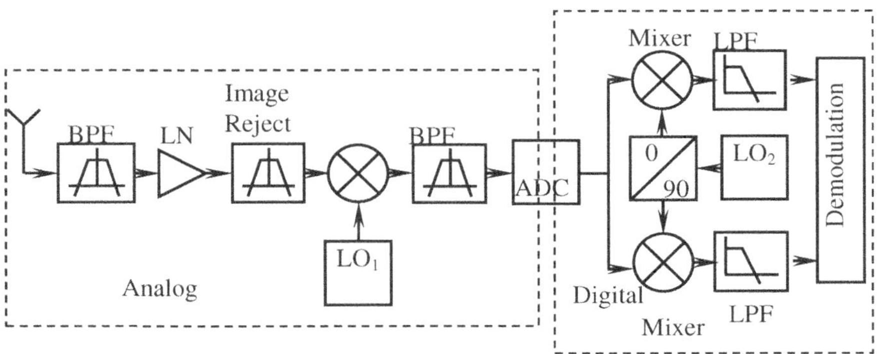

Fig. 3.52 Digital IF architecture

3.4.3.5 Digital – IF Receiver

One idea is to move the ADC converter after the first down-conversion but just before the second down-conversion. We obtain the structure of a digital – IF receiver, very similar with the heterodyne receiver architecture (fig. 3.52). In this case the second down-conversion and subsequent filtering can be done digitally. The principal problem here is the performance of the ADC and it is necessary to use a sufficiently low IF. But in this case is impossible to use band-pass filtering to suppress the image frequency (Weigel et al. 2001).

3.4.4 Fully Reconfigurable Analogue Filters Design

In (Stehr et al. 2003) is described a Fully Differential CMOS 4th Order reconfigurable gm-C Low Pass filter. This filter supports different modes including GSM, IS-95 and UMTS. The filter is a part of a DCR capable to receive 2G and 3G standards using the same receive path. The filters are placed between the mixer and ADC (figure 3.53) and are implemented in a standard 0.35 um CMOS process.

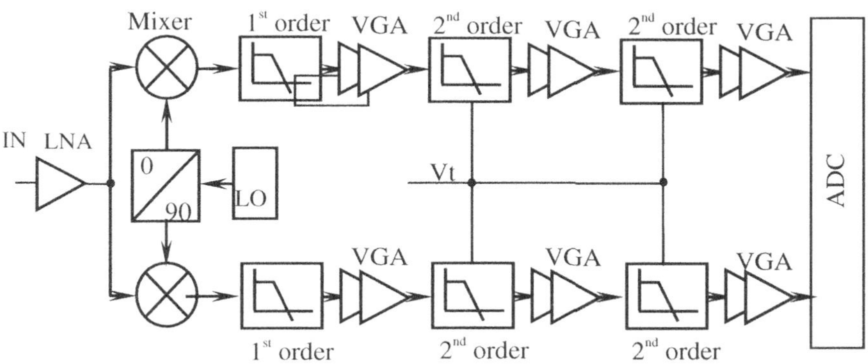

Fig. 3.53 Direct Conversion receiver block diagram

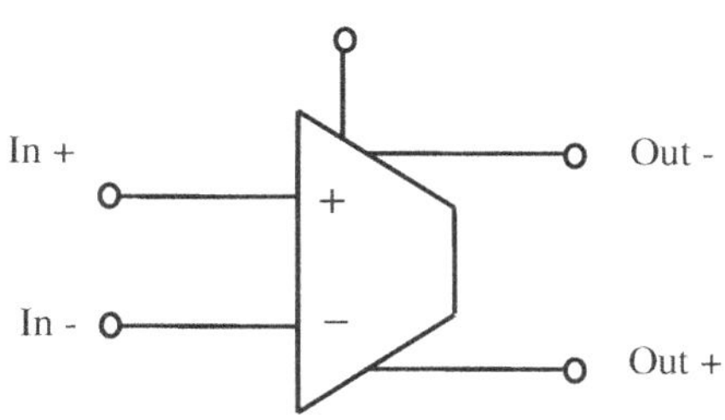

Fig. 3.54 Unit transconductance element

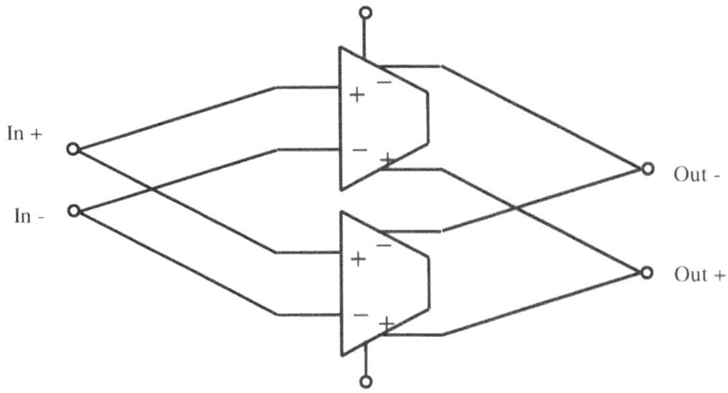

Fig. 3.55 Parallel connection of two unit transconductors

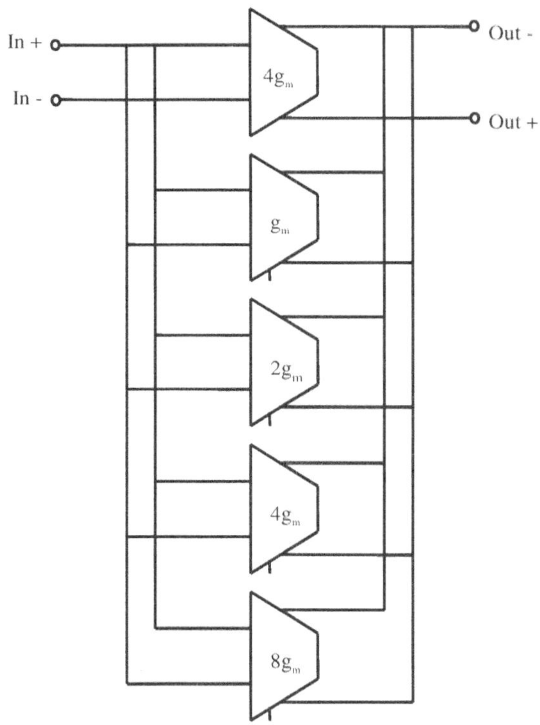

Fig. 3.56 Four-bit programmable transconductor

In (Pavan et al. 1999) is presented a programmable fourth – order Butterworth continuous–time filter with a bandwidth from 60-350 MHz, implemented in a 0.25 um digital CMOS process.

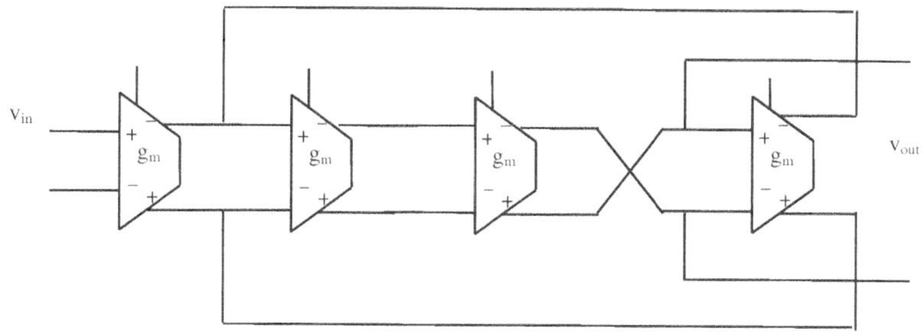

Fig. 3.57 Second order filter

The programmability is obtained using a unit transconductance element like in fig 3.54. This circuit has a transconductance equal with g_m if b=1 and equal with zero when b=0.

This could be seen as a programmable transconductor. We can consider the parallel connection of two identical unit transconductors as shown in figure 3.55. For two different values of b_1, we obtain an equivalent transconductance equal with g_m, or $2g_m$, depending on the b_1 value. We can see that the capacitances in all nodes are the same, not depending on the value of the control bit.

For a larger programmability range, the authors propose a solution with composite transconductors, each consisting of 19 unit transconductors as shown in figure 3.56. (Pavan et al. 1999). Based on the control word (b_0 b_1 b_2 b_3), the OTA can be configured as $4g_m$ when the control word is 0000, $5g_m$ when the control word is 0001, $6g_m$ when the control word is 0010, and $19g_m$ when the control word is 1111. The coarse programming range is between 19/4 and 3.75

The filter consists of two second order filter sections, each of them of the type depicted in figure 3.57.

3.4.5 *Reconfigurable Filter Stage for a Combined Zero-IF/Low-IF Radio Architecture*

3.4.5.1 Flexible Zero-IF/Low-IF Radio Architecture

A combined zero-IF/low-IF transceiver architecture is considered in (Hintea et al. 2007) as the architecture suitable for a highly adaptable and flexible radio. This approach is based on the existence of common functional elements between the direct conversion and the low-IF architectures.

In a zero-IF receiver the radio frequency (RF) signal band is down-converted directly to baseband by multiplication with a complex signal generated by the local oscillator (LO) at the same frequency with the RF carrier. The local oscillator signal is generated in quadrature to circumvent the image problem and to allow the demodulation of the useful signal. However, the difference in amplitude and phase between the quadrature signals I and Q determine the precision at which the image signal can be suppressed (Csipkes et al. 2003). Phase and amplitude errors between the two locally-generated orthogonal signals and the mismatch of the circuits on the two processing paths are difficult to overcome. Another problem that ultimately limits the implementation of this architecture is due to the parasitic baseband signals created during the down-conversion; they are mainly the result of crosstalk between the RF and the LO inputs of the mixer.

More recently, direct conversion implementations have been reported for wideband signals such as CDMA signals (Maurer et al. 2005). The inherently wide band of these signals (in this work we consider wideband the signals that have a bandwidth in the range of MHz, compared to narrow band signals with bandwidth in the range of kHz) makes them suitable to the direct conversion architecture. This approach was formulated in a number of publications (Steyaert et al. 1997; Tuttlebee 2002).

The low-IF receiver architecture is intended for narrow band signals such as the GSM 200kHz channel. The RF signal is down-converted at a sufficiently high frequency to allow separation from the baseband DC offsets, but at the same time sufficiently low to allow a good integrability and linearity of the circuits. In this paper the circuits have been designed using the G_m-C technique and an IF at 10 MHz has been chosen. Careful examination of the analogue front-end for the low-IF and zero-IF structures reveals that, apart from a different local oscillator frequency, the relevant difference between the two is in the channel filtering stage: a pair of low-pass filters is needed for the former and a band pass filter is required for the low-IF. Incorporating the constructive units of the two quadrature low-pass filters to construct a complex band-pass filter offers the advantage of reusing common hardware elements and leads to a highly adaptable circuit structure.

The proposed reconfigurable filter is highlighted in figure 3.58, integrated into a combined zero-IF/low-IF flexible receiver architecture. After a broad band-pass filtering (BPF) and some low noise amplification (LNA), the CDMA signal (over continuous line paths in figure 3.58) is mixed down to baseband (MIX) where the DC offsets are removed using two passive high-pass filters (HPF). A three stage amplifier follows (AMP) to allow a gain control of the signal. Only the quadrature

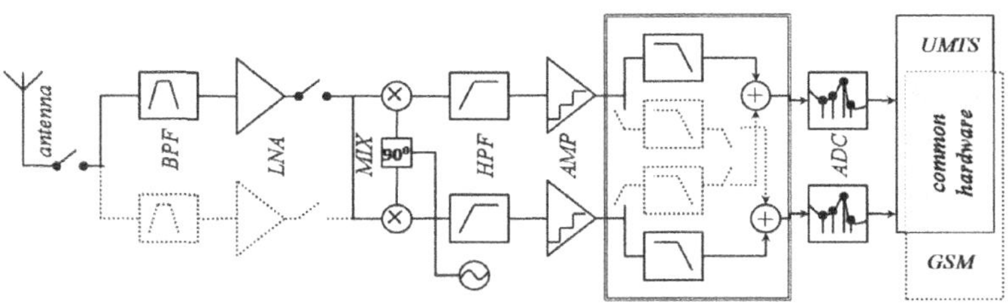

Fig. 3.58 The combined zero-IF/low-IF architecture

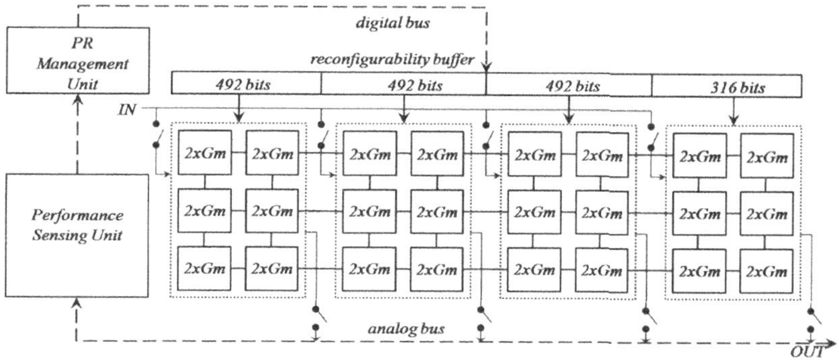

Fig. 3.59 A system view of the analog array

pair of low-pass filters is active for this configuration, performing the channel selection for the broadband signal. The processing stages for the narrow band GSM signals (over dotted line in figure 3.59) are similar, only that the central frequency translation units are connected to the quadrature paths to implement the complex band-pass filter. After being band-pass filtered at a central frequency of 10 MHz with a bandwidth of 2 MHz, the resulting signal is oversampled and further down-converted and filtered in baseband in the digital domain (Rus et al. 2007).

The receiver structure presented in figure 3.58 implements the concept of hardware reuse in the context of analogue circuits; this leads to a minimum of hardware redundancy, while only a few components close to the antenna are designed to work at high frequency. The low-pass real/band-pass complex analogue filters are implemented on an analogue array, discussed in what follows.

3.4.5.2 Transconductor-Based Reconfigurable and Programmable Analogue Array

At a principle level, the architecture of the analogue array is shown in figure 3.59. The system consists in a total of 48 programmable and reconfigurable transconductor units grouped in pairs, a Performance Sensing Unit, a Programmability and Re-configurability (PR) Management Unit, an analogue and a digital bus. The analogue bus connects the outputs from any G_m cell to the system output and at the same time feeds this signal to the Performance Sensing Unit that measures its deviation from the desired waveform; the PR Management Unit must be capable to ensure that appropriate control bits are fed into the array to obtain the desired output. It is envisioned for this control unit to be sufficiently intelligent to automatize the generation of the control bits according to a specific memorized algorithm such that the generated configuration makes use of the minimum possible array elements. The structure of one fully differential programmable and reconfigurable

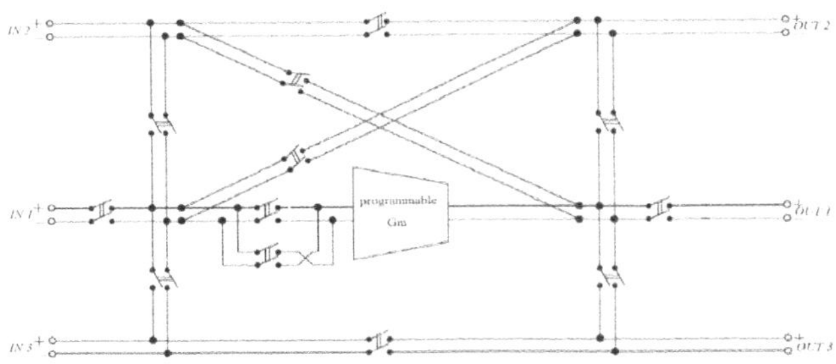

Fig. 3.60 The programmable and reconfigurable G_m cell

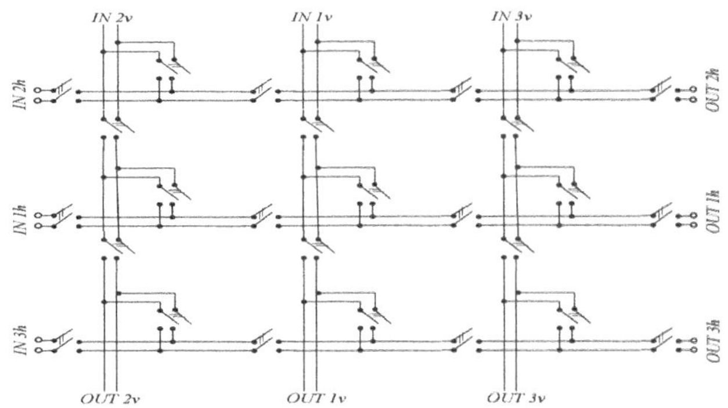

Fig. 3.61 The reconfigurable interconnection unit

Fig. 3.62 An odd order generalized passive ladder

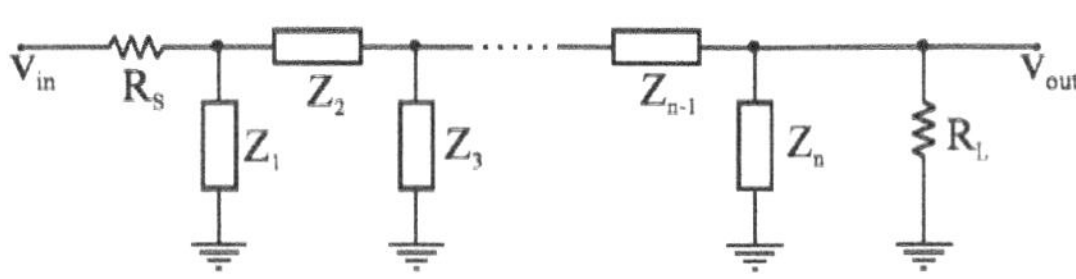

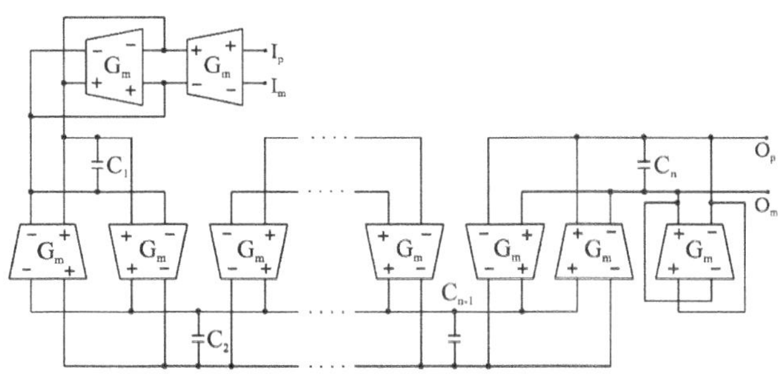

Fig. 3.63 Fully balanced odd order low pass filter

G_m cell is presented in Figure 3.60 (Rus et al. 2007). The switches ensure that the cell may simply pass the signal unaltered to other cells or route it to the inverting or non-inverting input of the G_m. Each two horizontally neighboring G_m cells in the analogue array are connected using the reconfigurable interconnection unit presented in Figure 3.61 This structure ensures that each of the three G_m cell outputs may connect to any of the three inputs from the neighboring cell. At the same time, the vertical inputs and outputs of the interconnection unit allow the achieving of connections between vertically neighboring G_m cells.

A number of 41 bits are necessary to control the configuration of each G_m cell and its associate interconnection unit. This leads to a total of approximately 2k reconfiguration bits which are stored in a buffer. The reconfigurable cell of Figure 3.61 may be configured to perform addition, subtraction, attenuation, integration or filtering (Pankiewicz et al. 2002). The 8x6 analogue array may implement various number of independent filters, depending on their hardware requirements and on the existence of corresponding independent input and output lines. For example, five real 4th order or three 7th order "leapfrog" low pass filters of various approximations may be implemented simultaneously, or twelve simultaneous biquadratic units. In section 3.4.5.4 the simulation results of implementing on the same array real low pass and complex band pass filters of orders 4, 5, 6 and 7 of Butterworth and Chebyshev approximations are presented. The low pass filters of orders 4, 5 and 6 have been implemented and simulated simultaneously.

3.4.5.3 Modular G_m-C State-Variable "Leapfrog" Filters

State variable filters stand out among several well known filter structures due to their low sensitivity performance, inherited from doubly terminated passive LC prototypes. Furthermore, the implementation based on operational transconductance amplifiers (OTA) allows the operation at high frequencies, making this filter type an ideal candidate for signal conditioning applications in various transceiver architectures.

Existing literature shows that, due to their highly modular structure, state variable ladder filters are particularly suitable for designing a generalized filter structure (Csipkes et al. 2003). The modularity is a prerequisite for the design of fully reconfigurable filters, as required by the SDR technology.

The versatility in operation is achieved by creating a circuit template that implements the signal flow graph of the generalized passive ladder prototype shown in figure 3.62. The impedances Z_i may be simple capacitances or more complicated structures involving generalized impedance converters (Csipkes et al. 2003).

Filters designed for communication systems exhibit mainly a low pass or band pass frequency response (Steyaert et al. 1997). In particular, when the imposed operating frequencies are not prohibitive and quadrature signal paths are available, band pass filters may also suppress the image signal in the intermediate frequency stage. In these cases the band pass response is obtained from the low pass transfer function by performing a linear frequency transformation (Steyaert et al. 1997).

The schematic of a fully differential odd order G_m-C low pass filter, corresponding to the prototype ladder in figure 3.62, is shown in figure 3.63 (Hintea et al. 2007).

It is of great importance that, when the capacitances have been set, the corner frequency of the low pass filter is proportional to the value of the transconductance. Consequently, the variation of the corner frequency is possible by simply adjusting G_m while maintaining capacitance values constant. This fact also leads to the idea of using the capacitance values to accommodate to different filter orders and topologies, without losing the desired approximation of the frequency response. Furthermore, the examination of the schematic in Figure 3.63 shows that the filter is built using functional modules that perform a signal addition, an impedance scaling and a current to voltage transformation. This module is built around two fully balanced trans-conductance amplifiers and a capacitance. A simple up-down cascade of identical modules allows the implementation of a filter with the desired order (Csipkes et al. 2003).

In the case of a low pass filter, the complete reconfiguration implies freely changing the frequency parameters, the filter order and the approximation while dynamically adjusting the current consumption. Considering these goals, it becomes very important to identify the impact of switching the filter order on the modular architecture of the circuit.

In the reconfiguration of the order, and implicitly of the topology, a very important role is played by the cells implementing the ladder termination on the load side of the filter. Figures 3.64 and 3.65 show the termination of two low pass filters of consecutive orders when n is odd number. The dashed lines show the components that must be disconnected when reconfiguring the filter order from n to n-1.

The comparison between the two circuits in Figures 3.64 and 3.65 suggests that the fundamental module should be implemented using switches that connect or disconnect the target transconductance cells depending on some digital control. Additionally, a negative feedback path must be created around the second OTA that effectively implements the load resistance of the ladder. The proposed schematic of the resulting fundamental module is shown in Figure 3.66. A simple cascade connection of identical modules, as many as required by the highest desired filter order, controlled by a decode logic, leads to fully reconfigurable low pass filter implementations.

When the filter order is decreased by one, the longitudinal switches S_{long} in the last module of the cascade are turned OFF, separating the unit from the rest of the ladder. Meanwhile, the switches S_{fb} are also turned OFF and the OTA cells enter in a power-down state. The new outputs of the filter will be the O_{phigh} and O_{mhigh} of the previous module. Furthermore, the switches S_{fb} of the previous module must be turned ON in order to shift the load resistance to the output of the lower order filter. The frequency parameters of the filter may be adjusted by means of variable transconductances, while the approximation is controlled by replacing all the capacitances in the circuit with programmable capacitor arrays. The programming algorithm of the frequency parameters is independent on the filter order and the necessary approximation.

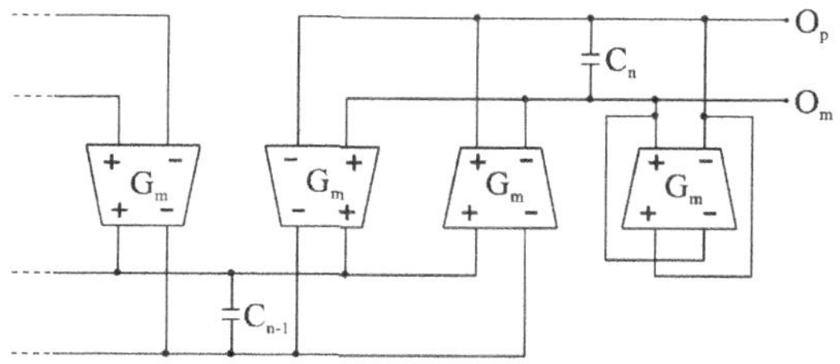

Fig. 3.64 The load side termination of an odd order OTA-C filter – nth order

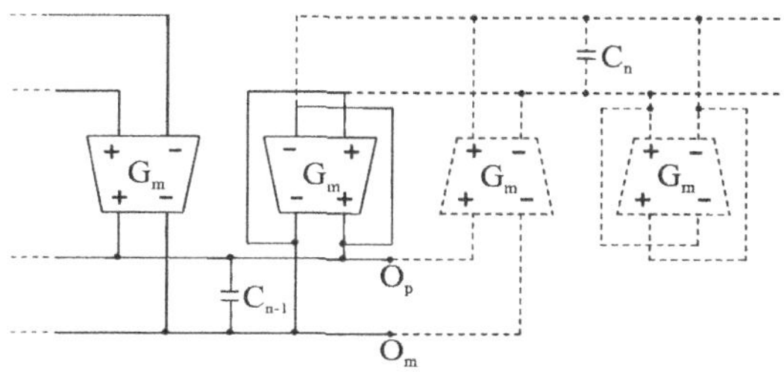

Fig. 3.65 The load side termination of an even order OTA-C filter – (n-1)th order

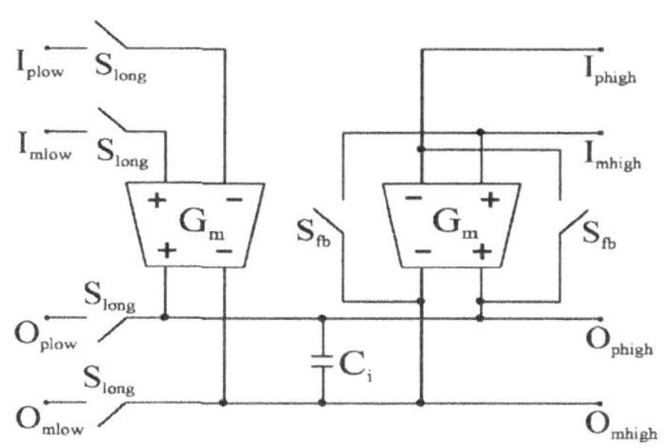

Fig. 3.66 The proposed elementary functional module of the reconfigurable low pass filter

The frequency response of a complex filter can be obtained from a low pass prototype by simply shifting the magnitude and phase responses to the desired centre frequency. This shift translates in shifting the impedances of all the reactive elements, capacitances in the OTA-C implementations, to the imposed centre frequency f_C. The frequency shift of the impedances is illustrated in equation (3.19) (Andreani et al. 2000; Steyaert et al. 1997).

$$j\omega C \rightarrow j(\omega \pm \omega_C)C = j\omega C \pm j\omega_C C \tag{3.19}$$

The changes in the complex admittance of a capacitor mean the addition of a quadrature signal to the current sunk by the capacitor. The implementation details are presented in Figure 3.67 (Hintea et al. 2007).

The complex input current may be written

$$i = i_I + ji_Q = v \cdot (sC + jG_{mf}) \tag{3.20}$$

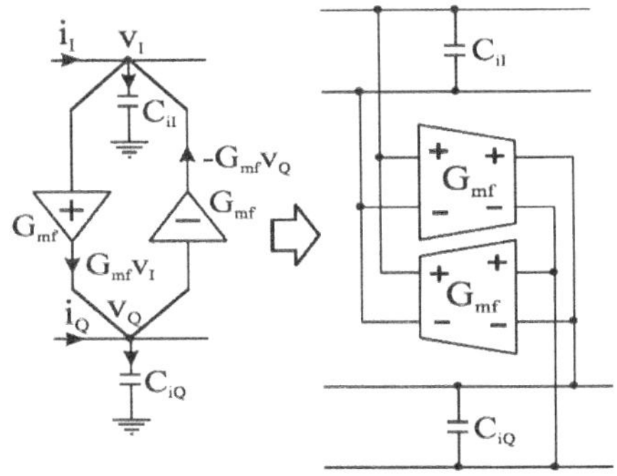

Fig. 3.67 Implementation of the linear frequency transformation and impedance shift

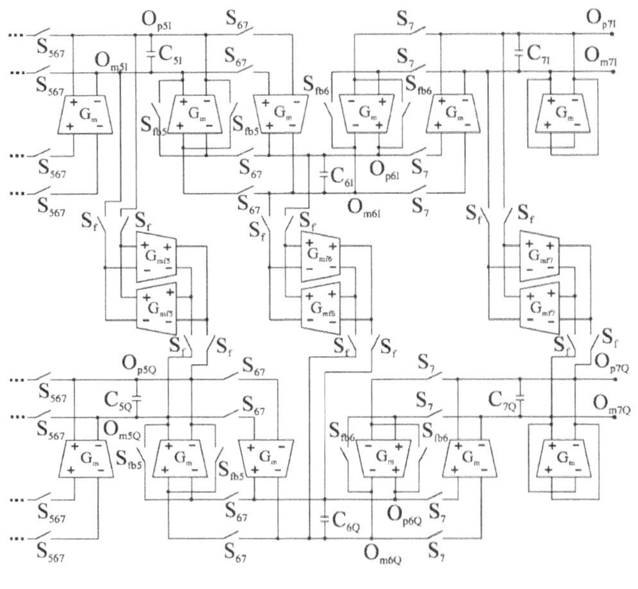

Fig. 3.68 Reconfigurable low pass/band pass filter section with the order switched between 5, 6 and 7

The admittance resulting from the equation (3.20) is then

$$Y_{IQ} = j\omega C + j\omega_C C = j\omega C + jG_{mf} \tag{3.21}$$

The resulting value for the transconductance cells that effectively shift the frequency response are then proportional to the capacitance and the desired centre frequency (Andreani et al. 2000; Csipkes et al. 2003).

The final structure of the complex filter can be obtained by simply duplicating the prototype low pass filter and connecting the bridges for all individual capacitor pairs. The described frequency transformation can be applied to any OTA-C low-pass filter.

Since the original goal was to implement a fully reconfigurable filter, the frequency shift units connecting the two low pass filters must be implemented with additional switches that allow the choice of low pass or complex band pass frequency response. A fully balanced reconfigurable low pass/band pass filter example is shown in figure 3.68. The switches are controlled according to Table 3.5.

Table 3.5 Switch states as functions of the order

Order	S_{567}	S_{67}	S_{7}	S_{fb5}	S_{fb6}	S_{fb7}
4	OFF	X	X	X	X	X
5	ON	OFF	X	ON	X	X
6	ON	ON	OFF	OFF	ON	X
7	ON	ON	ON	OFF	OFF	ON

The choice between low pass and band pass type transfer functions is determined by an additional bit that causes the switches S_f to be turned ON (band pass) or OFF (low pass).

3.4.5.4 Simulation Results

The transconductance amplifier model used to simulate the reconfigurable filter architecture emulates the operation of a typical folded cascode OTA. The circuit models the effects of the differential input stage bias current on the transconductance, the overdrive voltages of the input transistors, a parasitic high frequency pole and the finite output resistance.

Several filters of different orders and approximations have been selected for demonstrating the functionality of the analogue array. In a first instance, four different low pass filters were implemented, of orders 4, 5, 6 and 7; for the two even-order filters a Butterworth type approximation was selected, while for the

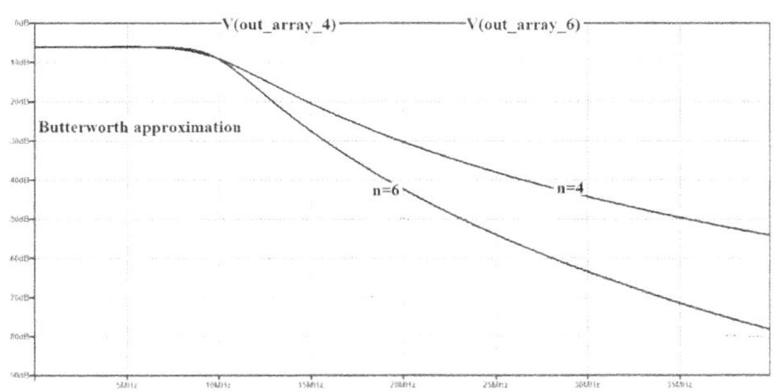

Fig. 3.69 The simulated magnitude response of the low pass filter for orders 4 and 6

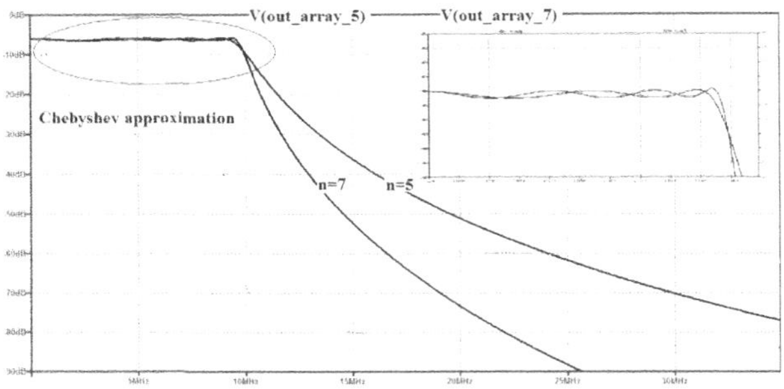

Fig. 3.70 The simulated magnitude response of the low pass filter for orders 5 and 7

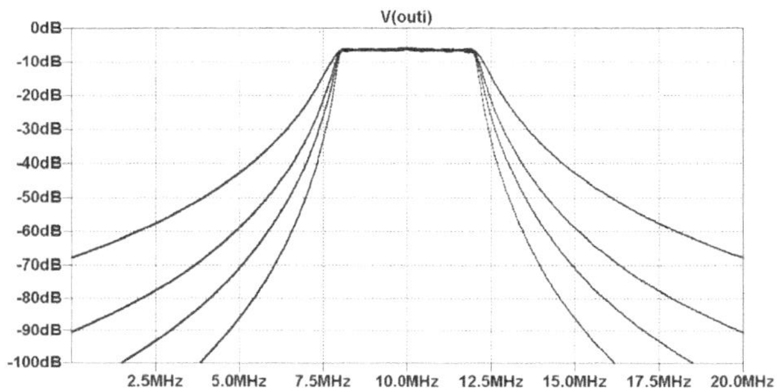

Fig. 3.71 The simulated magnitude response of the complex band pass filter for orders 4, 5, 6 and 7

odd-order ones a 0.5dB ripple Chebyshev type approximation. A 10MHz corner frequency was set in all cases. Any three out of the four filters may be implemented simultaneously on the same array. In this case the three lower order filters were implemented and simulated simultaneously.

The simulated magnitude response of the filter for orders 4, 5, 6 and 7 is presented in figures 3.69 and 3.70. Figure 3.71 presents the effects of sequentially changing the order of the complex band pass filter from 4 to 7. The bandwidth of the filter has been set at 4MHz at the centre frequency of 10MHz.

3.4.5.5 Conclusions

We described above the implementation of a low-IF filter stage in an SDR-enabled radio using a trans-conductor-based analogue array. As demonstration, the analogue array has been programmed to emulate several filter topologies, of different order and approximation, both in the real and in the complex domains. The simulated filters have been built around the synthesis method using state variables and the functional implementation of a passive ladder prototype. The target application of the proposed circuit is a versatile transceiver front end specifically intended for signal conditioning in a software defined radio environment.

The innate features of programmability and reconfigurability of the analogue array allow the implementation of low pass or complex band pass transfer functions of different orders and approximations.

The state-variable "leapfrog" synthesis method has been chosen for the implementation of the filters due to its inherent modularity; however, any synthesis method may be applied and the configurations of the resulting filter architectures downloaded as digital control into the array.

The main advantage of the analogue array circuit is its potential to accommodate different transceiver architectures (direct conversion with low pass and quadrature low-IF with band pass responses have been considered). Therefore, the performances in terms of spurious signal suppression and DC offset reduction may be optimized, depending on the spectral distribution of the processed signal.

Field Programmable Analog Array techniques are suitable for implementing circuits that require a digital control over their parameters and a wide interval of variation for these parameters; one such example illustrated here refers to the design of filters for programmable or multi-standard wireless transceivers where various frequency parameters are required to enable compatibility with a multitude of wireless standards.

3.4.6 Variable Gain Amplifiers

The Variable Gain Amplifier (VGA) is a circuit that automatically controls its gain in response to the amplitude of the input signal, leading to a constant-amplitude output. The (VGA) is normally employed in a feedback loop to implement an automatic gain control (AGC) amplifier. There are two approaches used to realize VGA's depending on whether the control signal is analogue or digital (Duong et al. 2006).

In digitally controlled VGA's gain varies as a discrete function of the control signal which can lead to discontinuous signal phases that can cause problems (Duong et al. 2006). In order to reduce the amount of jumps, a large number of control bits are required with digitally controlled VGA's.

In radio receivers, VGA is usually placed in front of ADC in order to maximize the dynamic range of the converter. In the case of direct conversion receiver, the baseband circuit controls the gain of the VGA with the help of a digital programming word delivered, in some cases, by the DSP.

In the case of direct conversion receiver the VGA must operate at low frequencies and could be built around operational amplifiers. One approach is to use a switching feedback resistor, as shown in figure 3.72.

Another solution is to vary the input resistor value as described in figure 3.73. In both architectures gain errors could appear due to the mismatches and variations of switching parameters with the temperature. But they have some important

Fig. 3.72 Gain variation using switching feedback resistor

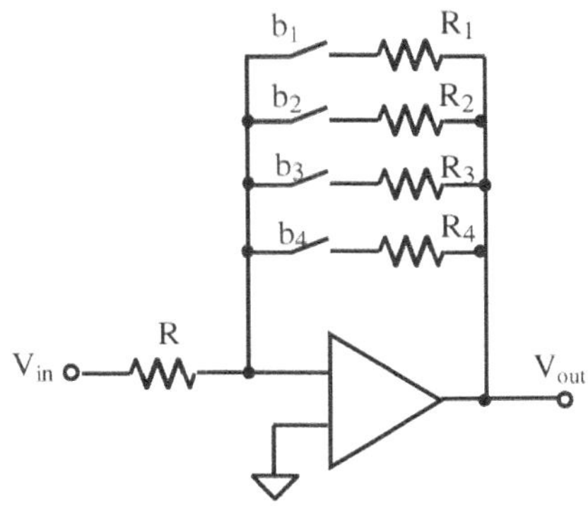

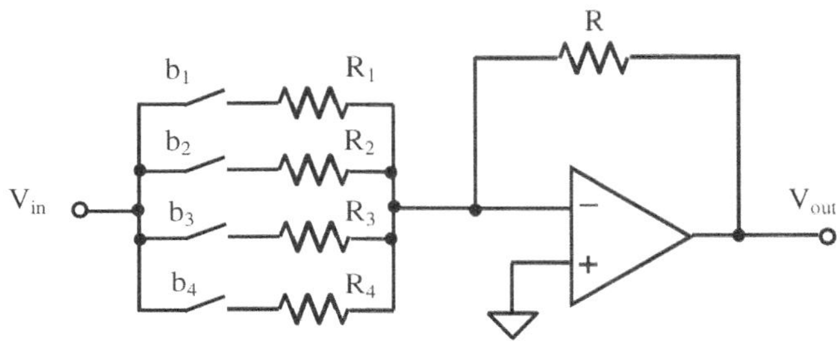

Fig. 3.73 Gain variation using switching input resistor

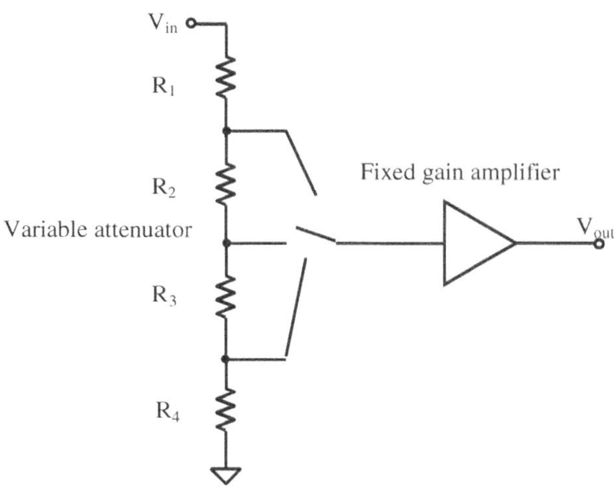

Fig. 3.74 Gain variation with a programmable attenuator and a fixed gain amplifier

advantages, such as good linearity and the capability to handle large signals (Csipkes 2006).

Another approach is depicted in figure 3.74 and consists of an amplifier with fixed gain and a variable attenuator. The total gain will be the product between the attenuation of the input network and the fixed gain of the amplifier. The principal advantage in this case is that it keeps the linearity by first reducing the input signal amplitude and makes possible to handle large input signals. But this solution has also some drawbacks. One of them is that the noise of the attenuation network is added directly to the desired signal. This means that the output resistance of the network must be kept at very small values and we need a buffer before the amplifier input.

The input signal is first attenuated and then amplified. But it is possible to invert the operations when the magnitude of the input signal is too small. In this case the amplifier is placed first and this will improve the noise performance.

In (Schelmbauer et al. 2000) is described a Programmable Gain Amplifier (PGA) based on a op-amp structure (figure 3.75). The different gain values are obtained by switching the feedback resistors and the resistors in the forward path. All the gain stages are in the range of 55.5 dB with a resolution of 0.5 dB.

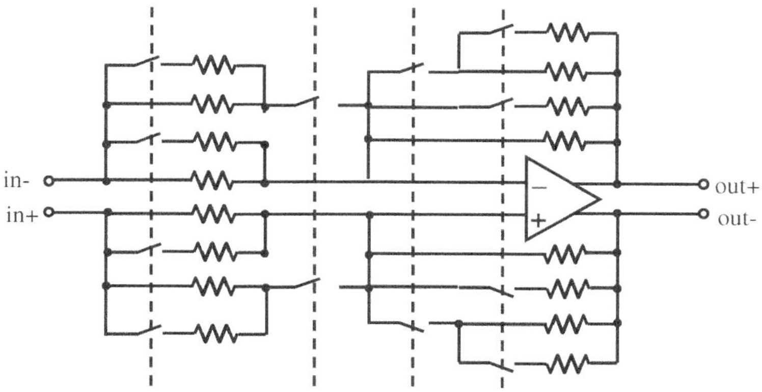

Fig. 3.75 PGA structure

In (Baez et al. 2004) is presented an analogue baseband block as a part of a DCR. The programmable gain amplifier (PGA) achieves gain from – 10dB to +50 dB in 5dB steps. There are two consecutive PGA stages, each of them consisting of an op amp with switched resistor network feedback (figure 3.76). Gain or attenuation can be achieved by changing the ratios R3/R1 and R4/R2. The variable values for resistors are obtained by using 4 or 7 switches disposed in parallel with

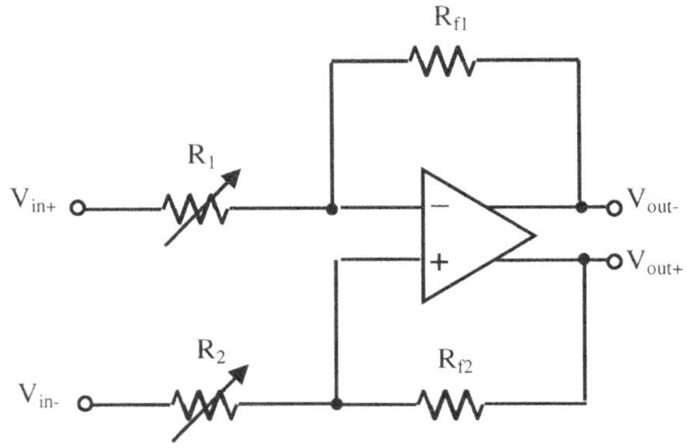

Fig. 3.76 Operational amplifier gain variation

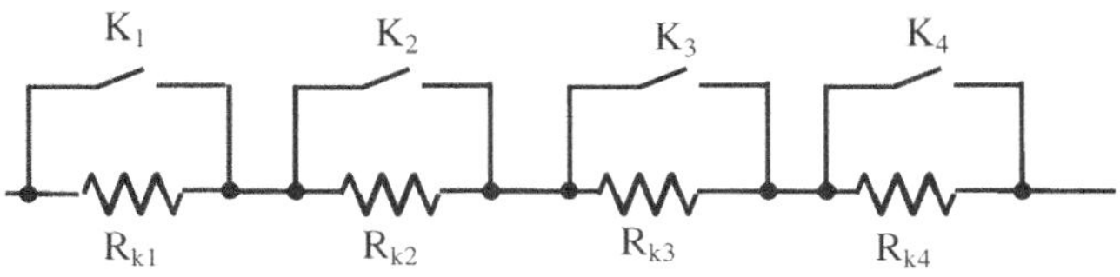

Fig. 3.77 Switching network

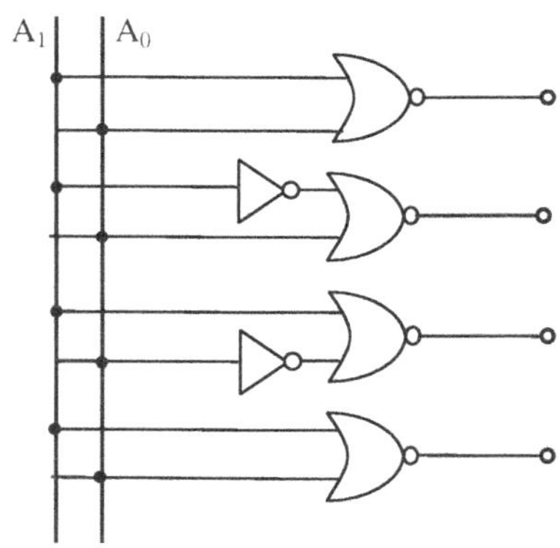

Fig. 3.78 Two bits decoder

the resistors (figure 3.77). The switches are turned on and off by a control signals provided by a decoder having 2 or 4 address bits (figure 3.78).

Another PGA configuration was introduced in (Wu et al. 2005) and it consists of 18 different individual amplifier cells connected in parallel and digitally controlled by a 5 to 18 de-multiplexer, as is depicted in figure 3.79. This circuit is capable to obtain a dynamic range of 51 dB with 3 dB gain control steps. There are 18 parallel amplifier cells, every cell having a different gain. The different values

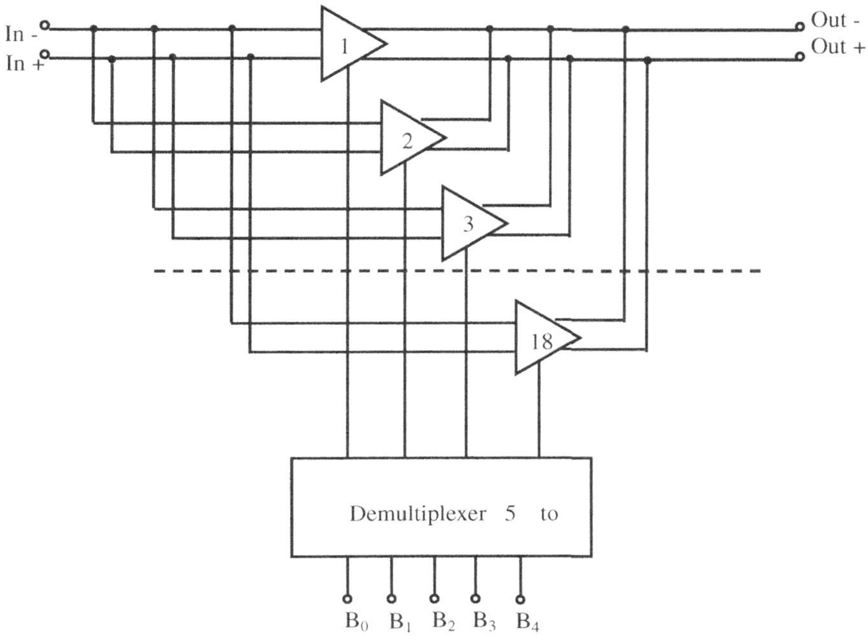

Fig. 3.79 Block diagram of the VGA

Fig. 3.80 Amplifier cell

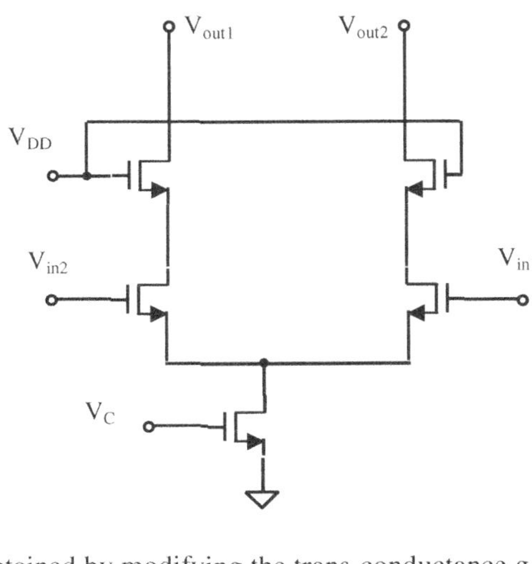

for the gain of MOS amplifier are obtained by modifying the trans-conductance g_m for every cell. There are used transistors with different gate width, as shown in figure 3.80.

3.4.7 Genetic Algorithms for Reconfigurable Analogue IF Filters Design

One of the critical problems in implementing reconfigurable analogue filters and variable gain amplifiers is that the values of manufactured analogue components are different from the designed specifications. The resistors and capacitors have values differing from the calculated ones. In this field of Intermediate Frequency (IF) filters and amplifiers, a very small deviation from the designed value could lead to unacceptable errors.

One good example how to use GA to design analogue integrated circuits was developed and described in (Murakawa et al. 1998). They present a new design method using GA calibration, the result being a chip fabricated in a 0.6 u, CMOS process (Murakawa et al. 2003), (Takahashi et al. 2003). The goal was to design an integrated Gm-C IF filter, with performances calibrated by GA.

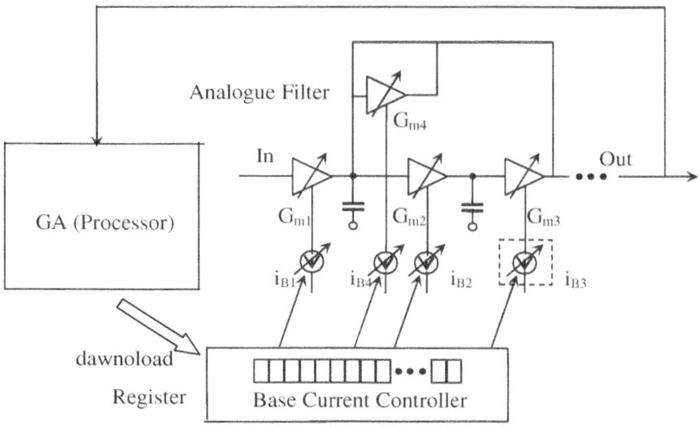

Fig. 3.81 The method using GA

This method provides some major advantages such as enhanced yield rates, smaller circuits and less power consumption. It was used to design a Gm-C IF filter consisting of 39 Gm amplifiers connected in cascade fashion (see figure 3.81). The transconductance of each amplifier is varied by altering the bias currents, which are digitally controlled with a bit string written to a register.

The designed filter is an 18th order linear filter consisting of 3 cascaded 6th order Gm-C leapfrog units as it is shown in figure 3.82.

The IF filter architecture is given in fig. 3.83. The master-slave technique is used to tune the Gm values. There are 39 programmable parameters in total: 19 for the central frequency, 18 for the bandwidth and the last 3 for filter gain.

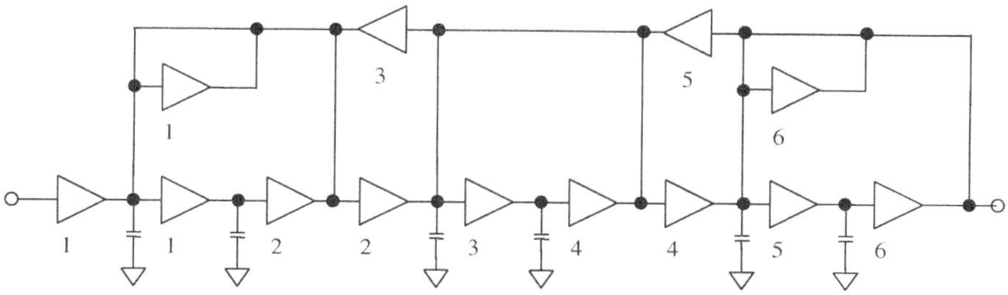

Fig. 3.82 Sixth order bandpass filter

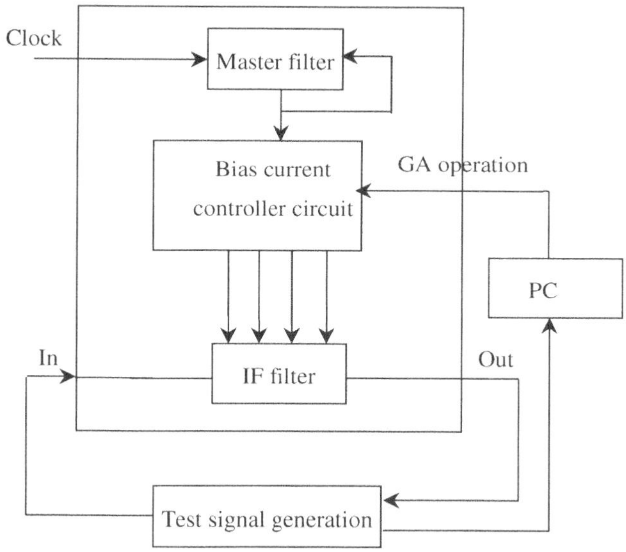

Fig. 3.83 IF filter architecture

The GA is used to calibrate the transconductance value of Gm amplifiers with a help of 39 chromosomes of 6 bits each. The 6 bits correspond to the switches in Digital Analogue Converter depicted in figure 3.84. The switches allow different combinations of currents values which fed the Gm amplifiers. The value I is set to $0.0039I_{ref}$.

The fitness function used for the filter design is:

$$\sum_{i=1}^{n} \omega_i \left| S_i(f_i) - O(f_i) \right|$$

where S is the ideal transfer function, O is the real circuit transfer function (Murakawa et al. 1998). The fitness function is the weighted sum of the deviations between the ideal gain $S(f_i)$ and the gain obtained $O(f_i)$ by the this chip at frequency f_i.

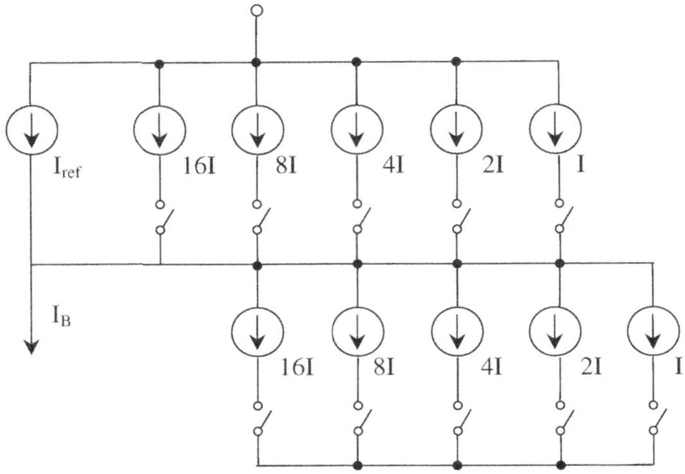

Fig. 3.84 DAC with bias currents controlled by GA

The GA calibration method is capable to optimize the frequency response and also to correct the group delay, that is impossible with the filter theory.

The results lead to 63% reduction in filter area and a 38% reduction in power dissipation. The explanation is the fact the amplifiers could be made very compact with a shortened channel width, reducing by this way the current dissipation. Using this method they obtain a yield rate of 97%.

3.5 Biomedical Engineering Applications

3.5.1 Electrical Stimulation and Neural Prosthesis

Many devices were designed in the last two decades in the neural prosthesis field. The main applications are the cochlea implant, foot stimulators for hemiplegic patients, stimulators for tetraplegics, phrenic nerve stimulation for respiration, the bladder stimulation and bowel management. The visual prostheses appear to be the most spectacular at this moment. Great efforts are being made in order to develop a totally implantable system within a human eye which can restore vision (Stieglitz et al. 2005).

Many other applications, such as pain management or stimulation of the limbs, are expected to be developed during the next years.

Also, Functional Electrical Stimulation (FES) system has been used for therapeutic purposes, not only for restoring activity in paralyzed members due to spinal cord injury or stroke (Sacristán et al. 2002).

New stimulating systems have been developed using different methods to deliver data transmission and power supply: from the totally external, continuing with those which use a cable connection between external and internal parts, until the implantable devices that use RF link to transfer wirelessly power and data (Sacristán et al. 2002).

The transcutaneous method has some important disadvantages such as the existence of a path for infection, and also the influence of the external noise leading to important interfering signals in the wires which carry the neural signals. (Harrison et al. 2007; Hintea 1996)

The wireless devices transfer information and power from an external primary transmitter coil to an implanted secondary receiver coil (see fig. 3.85).

Important efforts were made to improve the power- transmission efficiency and coupling insensitivity (Ko et al. 1977; Hintea 1986; Galbraith et al. 1987; Zierhofer and Hochmair 1990; Zierhofer and Hochmair 1996).

Neural prostheses are electronic devices meant to restore by electrical stimulation the function of nervous system, disabled or damaged from different causes.

This method is known as functional electrical (or neuromuscular) stimulation (FES or FNS) and it is used to generate action potentials applied to the nerve fibbers in order to produce muscles contractions. This will lead to a functional movement which is able to offer the patient a certain regain of control over the paralyzed muscles.

The implantable electronic circuits are not only complex circuits from the electronic point of view, but they have to follow some very strict conditions. One of them is the biocompatibility of all materials and their encapsulation which are surgically implanted in living tissues. These materials must be compatible with the tissues such that they won't influence the tissues function, by disrupting the activity or creating adverse response.

Another aspect is the manufacturing process which has to be made in controlled clean-room environments and followed by very complex testing procedures.

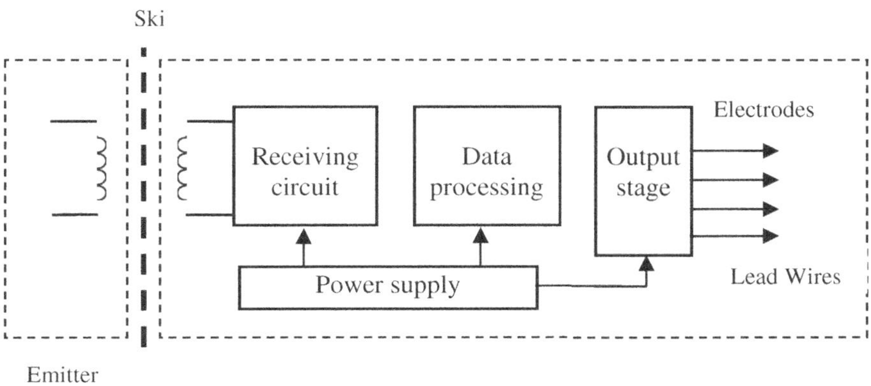

Fig. 3.85 Block diagram of implantable stimulator

A very important feature is the independence with the electromagnetic interference (EMI) and the elimination of any risk in the normal devices functionality (Strojnik et al. 2000).

Many difficulties appeared in the past because of encapsulation and connection problems or different situations of leaking or corrosion. One challenge in this field of research is also the complexity of processing data circuitry, the extension from single-channel to multi-channel stimulation, methods of multiplexing or parallel processing data. Another specific feature is that they need a lot of experimental work especially in order to find the best electric waveform needed to stimulate different cells.

The main problem of data transmission is the fact that tissues have high electromagnetic absorption at higher frequencies. This limits the bandwidth data transmission and it is difficult to obtain data rates above 1 Mbps (Dong et al. 2006). This could be a strong barrier in applications like visual prostheses which have not yet been widely utilized in the blind though extensive research has been performed. One of the major technological challenges of an implantable vision restoration device is to achieve a wide bandwidth with a carrier frequency that is limited to about 20 MHz (Zrenner 2002).

3.5.2 Cochlear Prosthesis via Telemetric Link

The implantable cochlear prosthesis is an electronic device implanted under the skin, used to stimulate the cochlear nerve by electrical signals. They are needed in those cases when the natural sensors are missing or damaged and must be bypassed the stimulation of cochlear nerve (Germanovix and Toumazou 2000). The electrodes are positioned in the cochlea and they stimulate the auditory nerve by electrical currents which create action potential transmitted to the brain. Present cochlear prostheses use a bundle of wire electrodes inserted into the spiral cochlea of the inner ear to electrically stimulate receptors in the auditory nerve, bypassing defective hair cells to create the perception of sound (Wang C S et al. 2005).

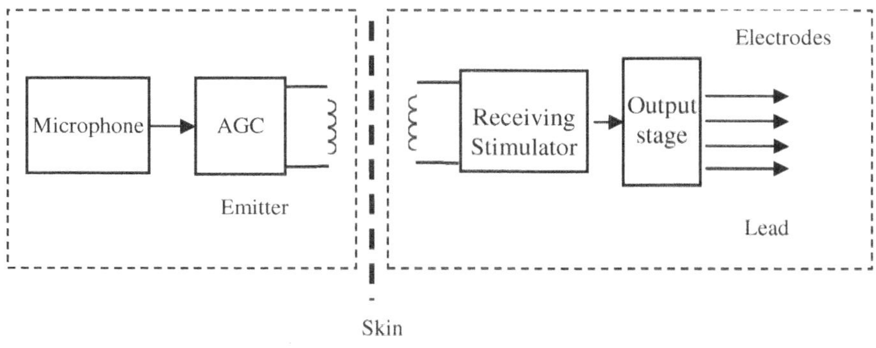

Fig. 3.86 Data transmission of a cochlear implant

The standard cochlear prosthesis has two parts: the internal one is fully implanted in the patient and consists of electrodes, receiver and stimulator. The external circuit contains the microphone, the amplifier, a signal processing and the transmitter. The system is depicted in fig. 3.86.

The microphone picks up the sounds from the environment and sends them to an amplifier. This amplifier has the role of compressing the signal in order for it to fit in the range between the minimum and maximum amplitude to the range between the electrical threshold and uncomfortable level of hearing (Germanovix and Toumazou 2000; Loizou 1998). The compressed signal is processed and transmitted to the internal part of the device.

The two parts are connected by an inductive transcutaneous RF link, which is capable to transmit both information and power for the internal circuits. The RF transmission works with the help of two coupled coils, one being external and the other internal. The receiver-stimulator is in fact a demodulator followed by circuits capable to deliver the necessary stimulating currents to a single electrode or to an array of electrodes.

3.5.3 Reconfigurable Circuits in Implantable Auditory Prosthesis

The information processing within an auditory prosthesis is very complex and could be made analogue or digital.

These devices, implantable or not, use a preamplifier with gain control which is present and necessary between the microphone and the ADC block. The main role is to compress the input signal, especially in order to prevent loud sounds to overload the ADC and prevent distortion.

This input compression limiter is in fact an automatic gain control (AGC) which could be seen as a reconfigurable circuit. The AGC design intended to be used in auditory prosthesis is described in different works as (Gata et al. 2002; Serdijn et al. 1994; Germanovix and Toumazou 2000).

The evolution of the totally external auditory prosthesis was in the direction of signals' processing's displacement from analogue towards digital. A typical example of processing the signals picked using a microphone, processed and rendered in a loudspeaker in a classic, without implant, auditory prosthesis is described in figure 3.87 (Gata et al. 2002). This application continues to be useful in those cases where the auditory natural chain is not interrupted, but it struggles with a strong diminishing of the amplification. This diminution of hearing compensates by using an electronic device which, beside the fact that it proper amplifies the signal supplied to a microphone, it also compresses the signal, different filtration on certain bands, until the application through a loudspeaker at the patient's tympanum surface.

All the parts of such a device are included in a single mixed-signal integrated circuit, except the microphone, the earphone and some capacitors. In this case an important part is also the AGC preamplifier including gain compression. This limits the amplitude provided by the microphone and prevents loud sounds from overloading the analogue-to-digital converter (ADC), this way preventing distortion (Gata et al. 2002).

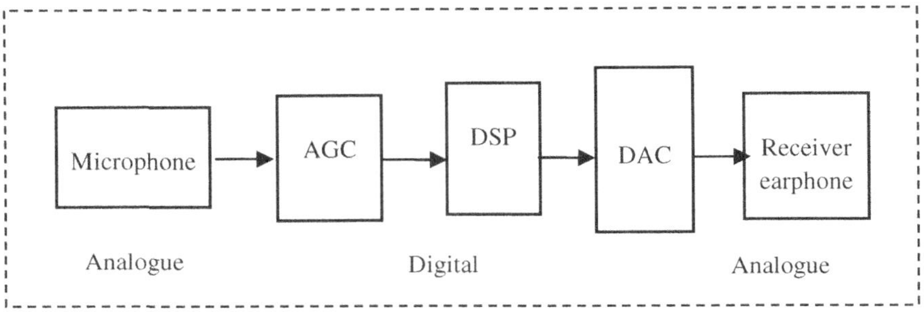

Fig. 3.87. Digital signal processing in a cochlear implant

The main advantages of digital signal processing are: it allows the adjusting of the stimulation signal's parameters through very flexible software algorithms. But there are many drawbacks as well such as: the high power consumption, the large volume of the device, high maintenance costs.

One solution would be the signals' analogical processing in a real analogical processor. In this situation several circuits from the processing chain need reconfigurability: amplifiers, band pass and low pass filters, tone adjusting circuits (Germanovix and Toumazou 2000).

One of the most difficult tasks is to separately adjust the stimulation levels of every channel. Many parameters are the object of this control such as the threshold and uncomfortable levels determined by the stimulation currents in every channel.

The analogue solution could lead to smaller size and low power consumption. They can be with single-channel or multi-channel stimulation.

The single channel devices stimulate a single place from the cochlea by using only one electrode. A widely known device is Vienna/3M single channel implant, whose block scheme is shown in fig. 3.88 (Loizou 1998).

Multiple channel implants ensure the stimulation in several places of the cochlea through an aria of electrodes. The electrodes are arranged in different places of

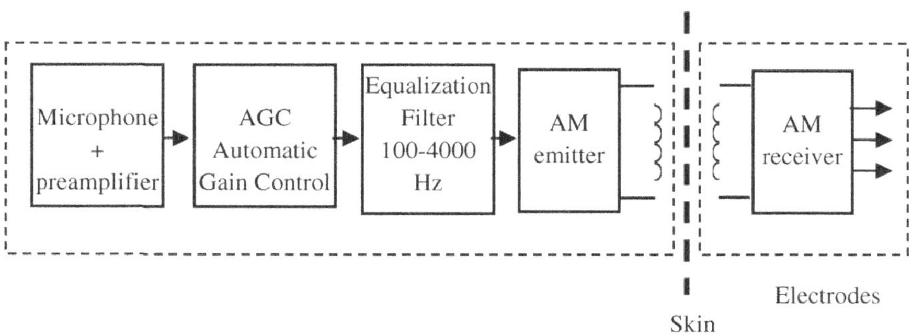

Fig. 3.88 Block diagram of a single channel stimulator

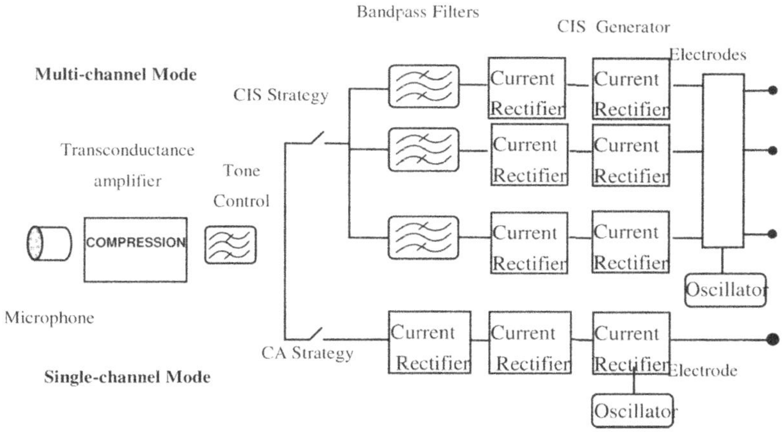

Fig. 3.89 Block diagram of a multi channel stimulator

the cochlea, susceptible at certain frequencies. This way, at the base of the cochlea higher frequencies are used, while towards the top lower ones are used.

The signal picked by the microphone is split into four or more bands, depending on the number of electrodes used (Germanovix and Toumazou 2000). The separation is done using band pass filters, which are able to cover the entire audio spectrum (see fig. 3.89).

As in the case of the single channel devices, there is a part which consists of a microphone, an amplifier with controlled gain and a tone controller, common to all channels. The amplifier has to compress the signal dynamically, between the minimum and maximum amplitude, such that it can be placed into the threshold domain and uncomfortable level of hearing.

In the case of the amplifiers with current output – transconductance, the voltage dynamics is transformed into current dynamics at the output using a control current. This way, we have a transconductance amplifier with controlled gain. In what concerns the dynamic acoustical domain, this is around 80 dB in case of a normal hearing and it can reach even 40 dB in a sensoneuronal hearing loss. The dynamic range for electrical stimulation of the diseased cochlea is around 10 dB.

(Germanovix and Toumazou 2000) introduced a tone control circuit capable to boost/cut off the low/high frequencies depending on the specific cochlea and environment.

3.5.4 AGCs in Auditory Prosthesis

One interesting approach and very suitable for digitally programmed AGC is described in (Gata et al. 2002) and the diagram is shown in figure 3.90. Here is used a resistances network capable to offer 83 gain steps of 0.5 dB from 1 to 40 dB. The resistor array switches are controlled by a bidirectional shift, which is able to change the state of only one switch per clock.

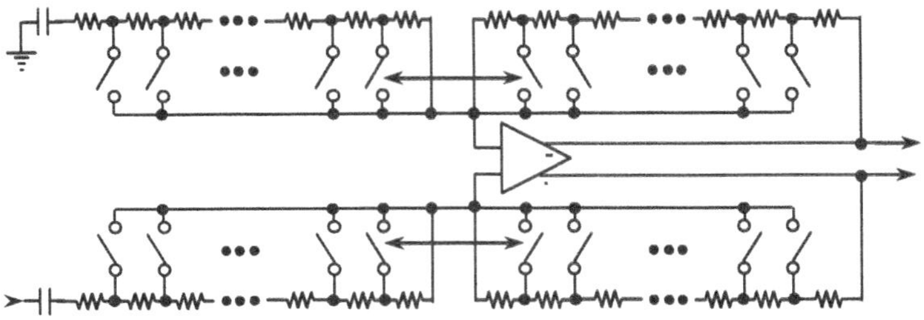

Fig. 3.90 Digitally controlled AGC

There are two main categories of AGCs: the input controlled and the output controlled AGC. The designed circuit is an output controlled with a compression range of 38 dB and a dynamic range of 62 dB with a bandwidth of minimum 100 kHz as described in fig. 3.91 (Serdijn et al. 1994).

Because the circuits in medical applications must work at low-voltage and low-power, it is preferred a scheme of an AGC operating in the current domain.

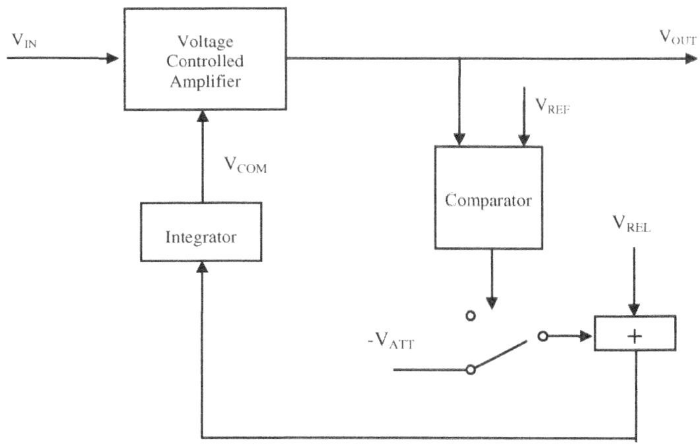

Fig. 3.91 AGC operating in the current domain

(Baker and Sarpeshkar 2006) present a complex approach of AGCs used in auditory prosthesis or bionic ear, containing single-loop and dual-loop variants. The devices have a dynamic range of 78 dB, a power consumption of less than 36 μW and an instantaneous dynamic range of operation of 58 dB. One important feature is that the dual-loop approach could be digitally programmed.

Amplifiers gain controlled (AGC) are frequently used within the bionic ear. One example is showed in fig. 3.92 and it was discussed in (Baker et al. 2006).

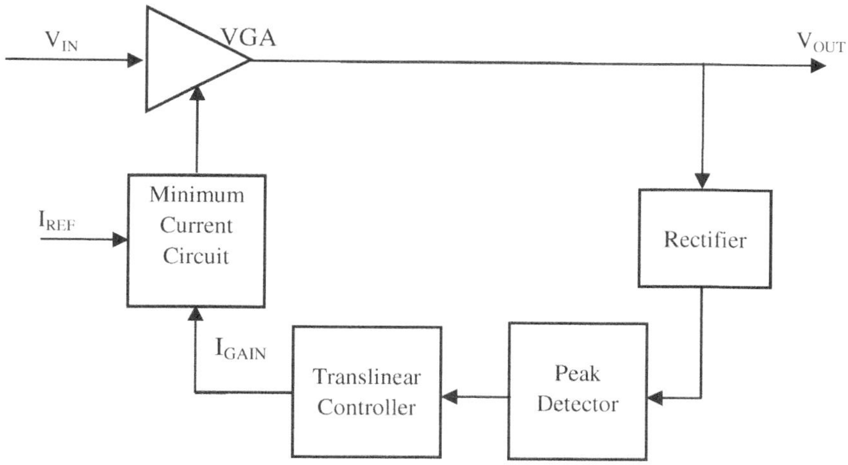

Fig. 3.92 VGA amplifier block diagram

The input is captured with the help of a VGA. The VGA's output is converted to a current and then rectified. The signal is applied to a trans-linear circuit which delivers a gain current compared with a reference one. The minimum-circuit controls the VGA's gain, comparing the two input currents.

The proposed structure of VGA has two wide-linear-range transconductors (WLRs) in a topology which uses an input voltage-to-current transconductor as a current source into a load transconductor.

Figure 3.93 shows the circuit scheme consisting of a voltage-to-current transconductor programmed by the current I_{GAIN} and a load transconductor programmed by current I_{REF} . The reference current could be digitally controlled.

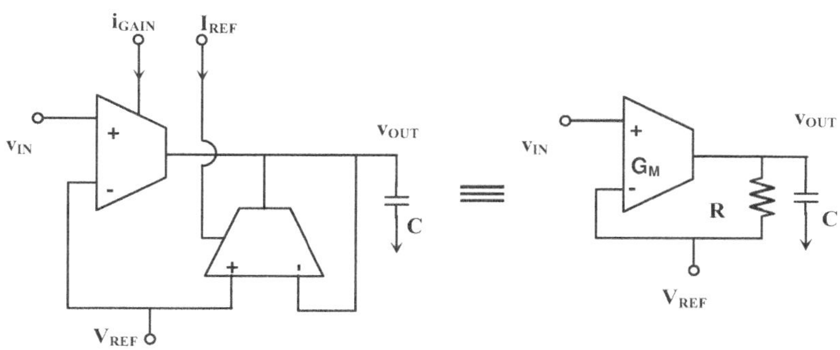

Fig. 3.93 VGA block diagram

3.5.5 *Binary Controlled Variable Gain Amplifiers*

3.5.5.1 Introduction

P*rogrammable* **G***ain* **A***mplifiers* (**PGA**) are important components in both analogue and digital auditory prosthesis. Their role is that of maintaining constant amplitude of the picked-up signals, while decreasing the influence of environment noise and other parasitic sources. These circuits are of great interest in many fields such as communication systems or neural networks and support vector machines. In particular, the programmable gain amplifiers digitally actioned can be seen as mixt signal multipliers, both analogue and digital.

The working manner of a programmable gain amplifier is shown in the functional block scheme belonging to figure 3.94.

Fig. 3.94 Programmable gain amplifier digitally controlled

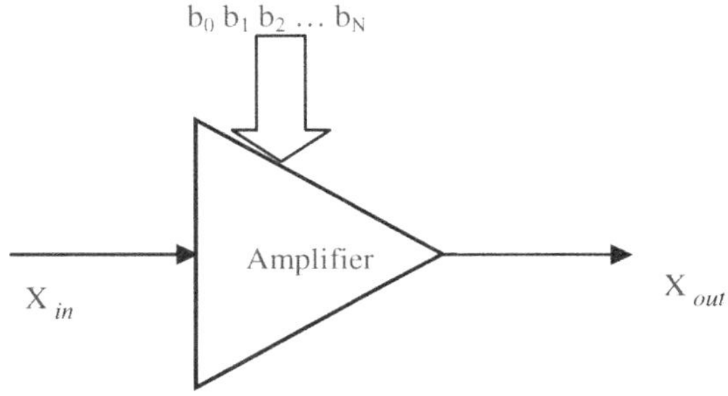

The general case is when the input (X_{in}) and output (X_{out}) signals can be both currents and voltages, and the gain G could be written as follows:

$$G = G_{\min} * \sum_{i=0}^{N} b_i 2^i \tag{3.22}$$

where $G_{\min}$ is the minimum control step, and b_i are the gain digital control bits (Festila et al. 2007). The sign could controlled, too.

3.5.5.2 Digitally Controlled Gain Amplifier with Current Mirrors

The simplest current amplifier, having a small or moderate gain, is the current mirror. The scheme of the circuit is given in fig. 3.95 (Allen et al. 1987; Wu 2003).

By writing the current-voltage dependence for both transistors and by also taking into account that the V_{GS} voltages are the same due to the circuit's topology, the current gain of the mirror will depend on the ratio between the geometrical parameters of the transistors, weighted by a coefficient depending on the drain-source voltages values.

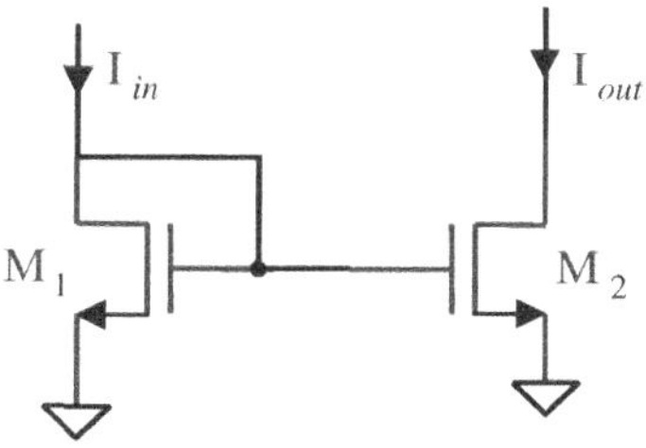

Fig. 3.95 Current amplifier implemented using a simple current mirror

$$G=-\frac{I_{out}}{I_{in}}=-\frac{W_2 L_2}{W_1 L_1}\cdot\left(\frac{V_{GS}-V_{Th2}}{V_{GS}-V_{Th1}}\right)^2\cdot\frac{1+\lambda V_{DS2}}{1+\lambda V_{DS1}}\cdot\frac{\mu_2 C_{ox2}}{\mu_1 C_{ox1}} \tag{3.23}$$

Where the measures have the following meanings:

- μ -the mobility of the charge carriers;
- C_{ox} - the specific capacity of the oxide found under the gate;
- W/L – the transistors' geometric parameters (width and length of the channel);
- V_{TH} - threshold voltage;
- λ - the modulation coefficient of the channel's length with the drain-source voltage.

In an ideal case we consider the threshold voltages and the trans-conductance coefficients μC_{ox} of both transistors as being equal and moreover, the products λV_{DS} can be considered a great deal smaller compared to the unity, and so the current amplification of the mirror is defined only by the ratio between the geometrical parameters (Allen et al. 1987):

$$G= -\frac{W_2 L_1}{W_1 L_2} \tag{3.24}$$

In reality, from the equation (3.23) it can be noticed that the disequilibrium between the drain-source voltages of the transistors, as well as the inherent parameters' variation caused by the manufacturing process have a significant influence on the value and precision of the gain.

The most used scheme, which allows the variation of the current gain, implies replacing M_2 transistor by a weighted aria of transistors, which through an adequate commutation provides the imposed values of the reflexion factor associated to the current mirror. This principle is depicted in Figure 3.96 (Fang 1995; Loh et al. 1992).

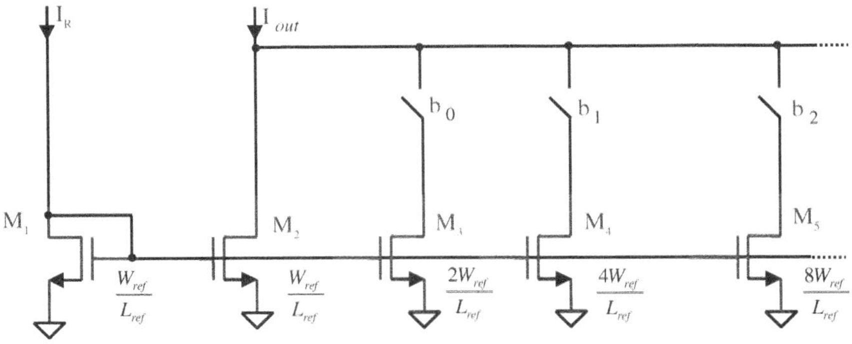

Fig. 3.96 The implementation principle of the current amplifier having variable gain and binary weightings

In this case, the ratios $\frac{W}{L}$ of the transistors are binary weighted such that the gain, which is always above unit, adapts the equation below:

$$G=\frac{I_{out}}{I_{in}}=-(1+\sum_{i=0}^{N-1} b_i 2^i) \tag{3.25}$$

The structure also allows the implementation of some below unit gains by properly modifying the W/L ratio, associated to the M_1 transistor.

The disadvantage of such a current amplifier structure is the lack of precision, generated mainly by the systematic error, introduced by the disequilibrium of the transistors' drain-source voltages from the output branch towards M_1 transistor. At the same time, the useful band of the amplifier is reduced by accentuating the input capacity simultaneously with the gain's increase. This requires using a larger number of parallel transistors and, implicitly, a larger area occupied by them.

The frequency response and the amplifier's precision implemented using a single current mirror can be improved by using some structures of cascode type (Allen et al. 1987; Wu 2003). The cascode reduces the zero effect over the frequency's response by decreasing the equivalent resistance seen by M_{2a} transistor in its drain.

This way, the gain of this transistor is decreased and, implicitly, the Miller effect introduced by the parasite grid-drain capacity M_{2a} is diminished. One must carefully take into account the geometry of the transistors which implement the switches. In order to equilibrate the equivalent series resistance of the switches on different branches, the W/L ratios of these transistors must be scaled according tot

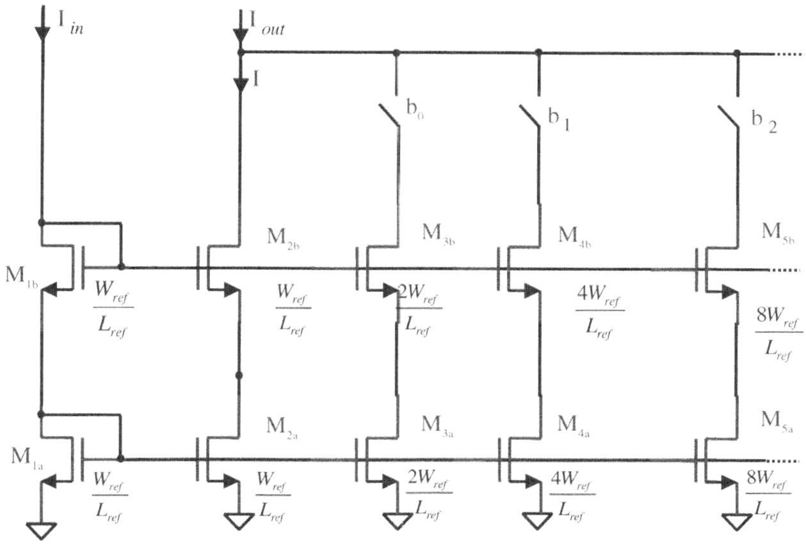

Fig. 3.97 The scheme of the current cascade amplifier with digitally controlled weights

the current which crosses them. The amplifier's main scheme having a negative variable gain is given in figure 3.97.

The gain has the same expression given by the equation (3.25), while the gain errors are diminished due to the balancing of the drain-source voltages of M_a transistors.

The circuit's disadvantage is the high relative minimum voltage not only at the output, but also at the input. This voltage is required in order to assure the desired polarisation and to keep the transistors from the output level of the current mirror having variable gain into a saturated regime.

An improved version of the cascode amplifier with variable gain can be obtained by using a cascode mirror of low voltage. The main scheme of the amplifier with variable gain is given in the figure 3.98 (Loh et al. 1992).

The branch containing M_{0a} and M_{0b} transistors has the role of generating the polarization voltage V_{casn}.

If the dimensions of M_{0b} are chosen identical to the ones of M_b transistors, then the voltage drop between the drain and source of M_{0a} will be approximately equal to the drain-source voltage chosen for M_a transistors.

The minimum voltage at the output of the circuit is reduced to V_{Th2b} in comparison to the classic cascode variant (Allen et al. 1987; Wu 2003). This way, the amplifier is suitable for its use in low voltage applications.

The frequency performances of this circuit and the gain's selection precision are similar to the ones of the classic cascode structure. The circuit can be used as a current amplifier in class A if the proper current sources are added, which fix the static functioning points of the programmable mirror transistors.

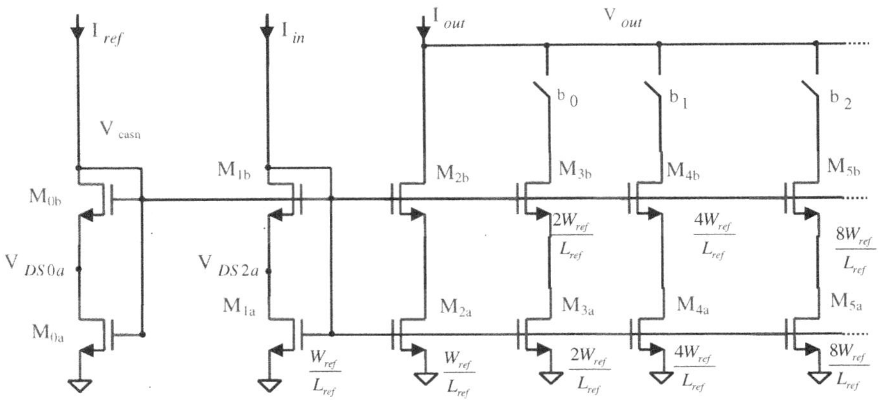

Fig. 3.98 Low voltage CASCODE current amplifier digitally controlled weights

3.5.5.3 Current Division Network

The current's division network (CDN) is based on the principle of R-2R resistances network used in numerical-analogue convertors and also in the amplifiers having different gain steps (multiple of 2). This architecture uses binary weighted current sources which are designed using identical MOS transistors. The division networks are widespread in the domain of numerical-digital conversion, especially due to the high working speed. The disadvantage appears mainly because of the large consumption of occupied surface on the chip due to the high number of needed sources, which increases exponentially with the number of bits of the D/A convertor. Moreover, another disadvantage is the difficulty of pairing the transistors which determine the output from the desired working parameters. In this case are necessary both a complex layout and a systematic distribution of these current sources which form the most significant bits. Another solution would be an auto-calibration circuit such as a dynamically pairing element, but this implies an even larger consumption both of surface and power.

The working principle scheme of an amplifier based on a current division network using an R-2R network is given in figure 3.99.

The R-2R network has a number of cells similar to that showed in fig. 3.100.

If we replace into this configuration the resistances by switches and MOS transistors we obtain an R-2R cell made only with MOS transistors, a cell which is shown in figure 3.101.

While the R resistance from the scale of the current division scheme is being replaced by T_2 transistor, T_1 and T_3 transistors (or T_4) work like the 2R part from the resistive equivalent circuit.

The current at the cell's input is split into two equal currents (only in the case when the W/L ratios are identical for all transistors), one of them being found in T_2 transistor's source having the half value of the input current's one, while the other current is being commuted either on the T_3 transistor's source or on T_4 's one.

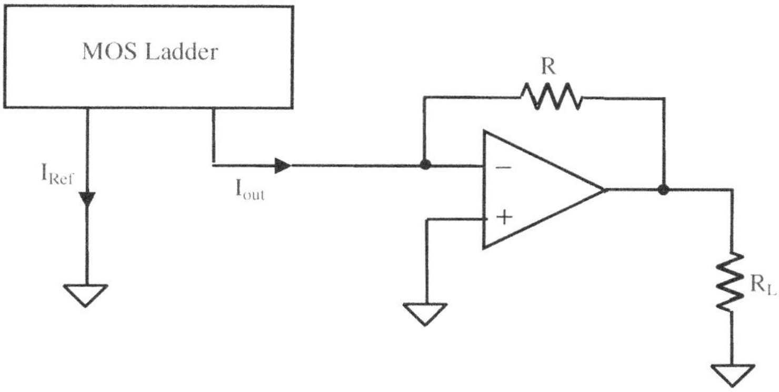

Fig. 3.99 Block scheme of the current division network

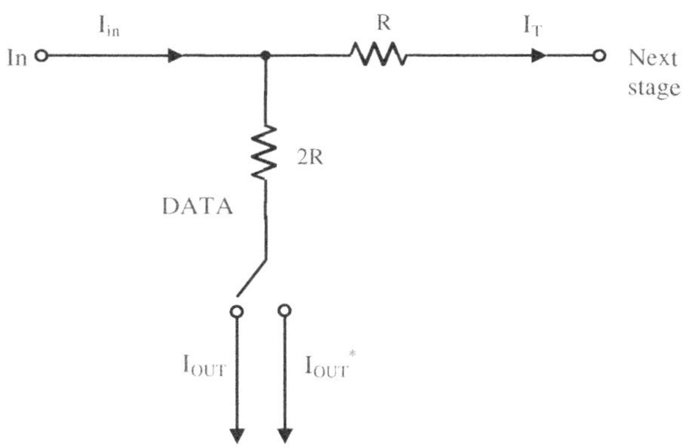

Fig. 3.100 Fragment from an R-2R network

The relation between the ouput currents of such a cell is:

$$I_{out}^{*} = I_{in} - \alpha \cdot I_{out}$$

where α is the logic control signal. The amplifying will depend on the logic control signal as it follows:

$$A = \frac{I_{out}}{I_{in}} = \frac{1}{\alpha}\left(1 - \frac{I_{out}^{*}}{I_{out}}\right)$$

Starting with the previous example a network designed only with MOS transistors can be achieved, as shown in figure 3.102 (Hammerschmied 2000). The scheme contains several cells in series and it is polarized through an I_{REF} current source, whose value determines the whole domain of the output current from the scale.

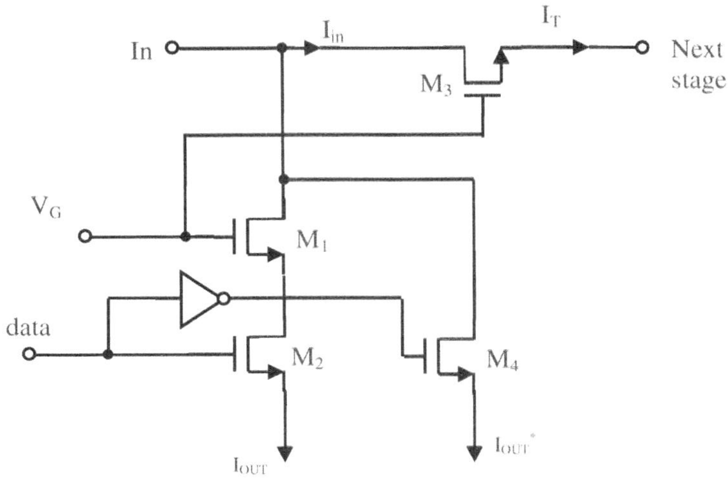

Fig. 3.101 A cell of R-2R network made by using only MOS transistors

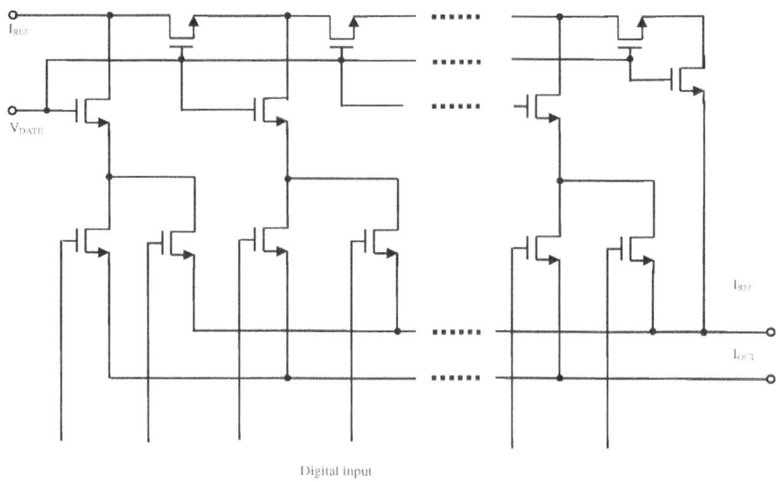

Fig. 3.102 Cascaded network containing several current division cells

3.5.5.4 Programmable Amplifiers with CDN

When we want to obtain an amplifier having variable gain we can use a resistive network followed by an AO which transforms the current into voltage. This kind of networks is widely used in DAC or ADC converters, but also in programmable amplifiers and filters.

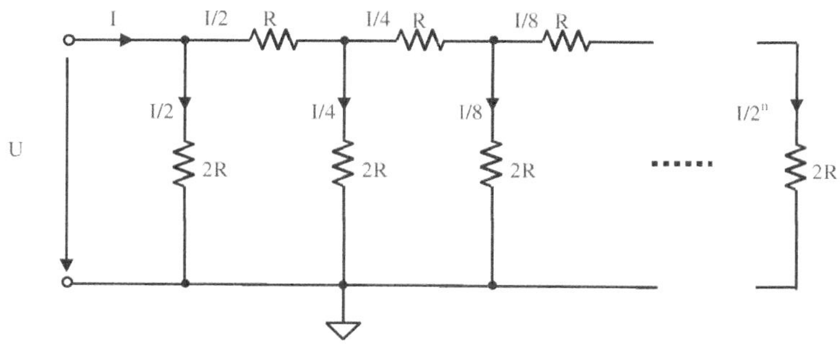

Fig. 3.103 R-2R Ladder

The scale resistive network has its electrical scheme presented in the figure 3.103. Certain conditions are imposed to it, such as the resistance in each node in the right should be R and the current division ratio from each node should be n.

If the condition that in each node the injected current must be divided by 2 is imposed (n = 1) it results that R_1 = 2R and R_2 = R, this way being obtained the R-2R resistive network.

The property of this network, that of dividing by two the current which enters each node and which is used in building the numerical-analogue convertors using R-2R network, is extremely useful at the step control of an amplifier's gain, such as the one in figure 3.104. Here AO has the role of transforming the current into a digitally controlled analogical voltage.

K_i switches, led by logical signals and made of MOS transistors, have the role of guiding the currents from the vertical braches towards the analogical adder or the 2R resistances' mass.

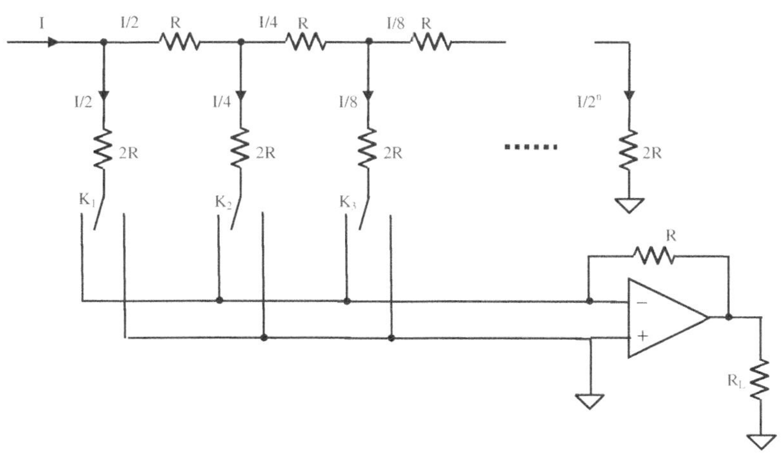

Fig. 3.104 AO R-2R Ladder

The currents applied to the K_i switches can be determined by taking into account that in each of the nodes in the R-2R network takes place a division of the incident current into two equal parts, such that:

$$\begin{cases} I_0 = -\dfrac{U_{ref}}{R} \\ I_1 = 2^{-1} I_0 \\ I_2 = 2^{-1} I_1 = 2^{-2} I_0 \\ \ldots\ldots\ldots \\ I_n = 2^{-n} I_0 \end{cases}$$

The current I can be obtained by summing all the currents afferent to the switches which are in position 1, so:

$$I = \sum_{i=1}^{n} I_i K_i = -\frac{U_{ref}}{R} \sum_{i=1}^{n} K_i 2^{-i}$$

From here also results the output voltage:

$$U_{out} = -RI$$

The value of the output voltage can be easily deduced using the fact that I represents the sum of the n currents added into the R resistance from the reaction:

$$Uout = R\left(\frac{K_1}{2}\frac{U_{ref}}{R} + \frac{K_2}{2^2}\frac{U_{ref}}{R} + \ldots + \frac{K_n}{2^n}\frac{U_{ref}}{R}\right) = U_{ref} \sum_{i=1}^{n} K_i 2^{-i}$$

This way, the value of the gain will have the following form:

$$A = \frac{U_{out}}{U_{ref}} = \sum_{i=1}^{n} K_i 2^{-i}$$

The meaning of this formula is the following: the amplifier's gain depends on the position of K_i switches (if K_i is on position 1 then the signal at that point will influence the amplifying factor with the weight 2^{-i}, and if K_i is on position 0 then the signal at that point won't be taken into account) as well as on the weights' value, precisely on the value of 2^{-i}.

The accuracy achieved by this converter is superior to other CNA variants because it uses only two values of R and 2R resistances, while the switches connected at low potential go in series with the same 2R resistance. In order to achieve high working speeds it is required that the network's resistances to be of small value.

3.5.5.5 Digitally Controlled Current Attenuator

The division of current networks could be very useful in various kinds of analogue signal processing, such as active filter with operational amplifiers and RC networks where we need to control the cutting frequency. Other examples are the controllable attenuation and A-to-D or D-to-A converters. MOS transistors are used as switches or amplifying elements replacing resistors or capacitors.

One useful application is the digitally control attenuator which is a sub unitary amplification circuit presented by (Bult and Geelen 1992). The attenuator is used to design a volume control circuit with the help of some cascaded attenuation cells. Half of them have a 12 dB attenuation per section and the other half achieve the fine steps of 2 dB per section.

The attenuation sections use a resistive R-2R ladder configuration as it is described in fig 3.105. There are two cells with two -6-dB ladder elements per section, thus resulting (in) a total attenuation of 12.04 dB which is transmitted to the next stage. The attenuation is digitally controlled, so when the digital input is high, the transistors are switched off. In this case the input is transmitted directly to the output without any attenuation.

When the digital input is low, then the transfer transistor is turned off and the attenuation section is switched on. The input current is now attenuated 12 dB and fed to the next section.

Starting from the cell presented in fig. 3.105, there are various possibilities to obtain different attenuations. In figure 3.106 is shown the fine section which is a R/5-4 ladder, leading to an attenuation of about 2 dB.

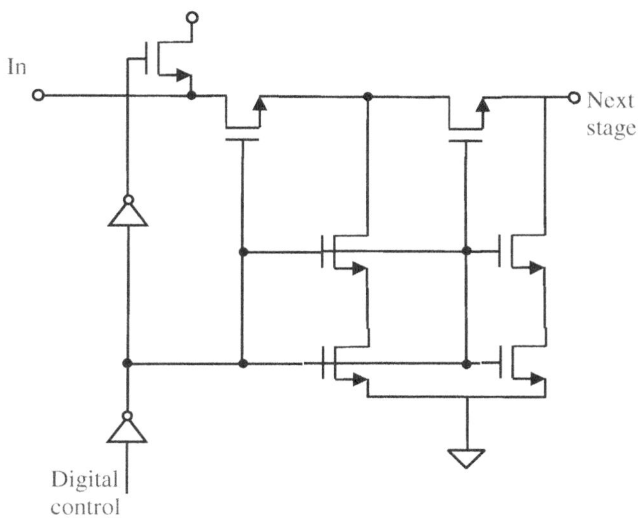

Fig. 3.105 MOS 12 dB attenuator cell

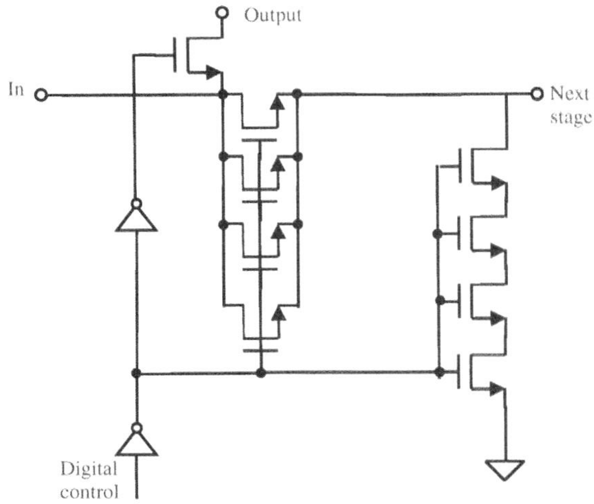

Fig. 3.106 The 2 dB section attenuator

This I-I convert could be completed with AOs, both at input and output and so results a V-V circuit as it is depicted in figure 3.107.

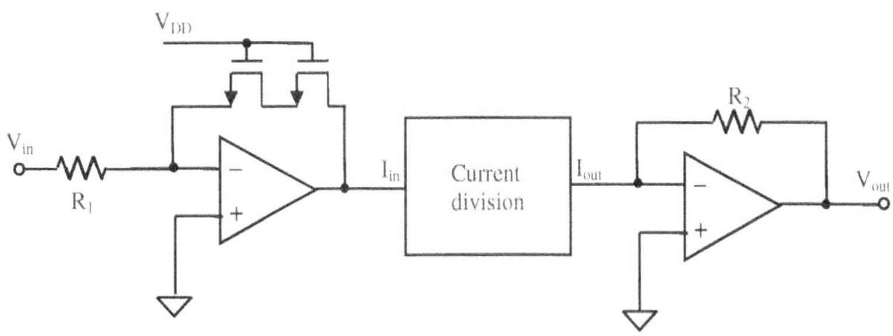

Fig. 3.107 Voltage to voltage attenuator

Sometimes we need a sub unitary amplification and in this case we can also digitally control the circuit response. The attenuator works on the principle of current's successive division by using a network of linearly polarized transistors.

The circuit described in figure 3.107 uses a dividing network similar to the R-2R scale used in DACs (Tarim et al. 2001; Pun et al. 2005). This could be useful in active filter with operational amplifiers and RC networks where we need to control the cutting frequency (Festila et al. 2007).

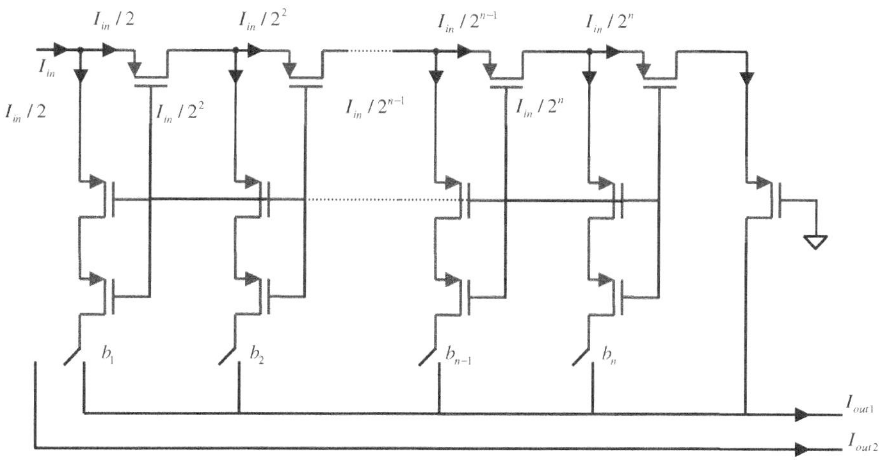

Fig. 3.108 Digitally controlled attenuator

The relations describing the dependence of output currents I_{out1} and I_{out2} on the input current I_{in} and the control bits b_i are given below:

$$\begin{cases} I_{out1} = I_{in} \cdot \left(\frac{1}{2^n} + \sum_{i=1}^{n} \frac{b_i}{2^i}\right) \\ I_{out2} = I_{in} \cdot \left(\frac{2^n - 1}{2^n} + \sum_{i=1}^{n} \frac{b_i}{2^i}\right) \end{cases}$$

The way of composing the total attenuation is given in table 3.6 for all configurations of the 3 control bits (Festila et al. 2007).

Table 3.6 The attenuation composition for 3 control bits

$b_1b_2b_3$	Attenuation composition	Total gain
000	1/8	0,125
001	1/8+1/8	0,250
010	1/8+1/4	0,375
011	1/8+1/8+1/8	0,500
100	1/8+1/2	0,625
101	1/8+1/8+1/2	0,750
110	1/8+1/2+1/4	0,875
111	1/8+1/8+1/4+1/2	1,000

3.6 Concluding Remarks

The proposed genetic algorithms applications in the field of electronic circuits design show a wide range of both analogue and digital possible utilizations. Evolutionary methods could be used in order to design analogue and digital structures. The concept of E*volvable* H*ardware* applied in adaptive systems could increase in the future, being very useful in telecommunications and medical applications.

One important feature of evolutionary methods applied in digital design is that they are different from the conventional designs. This explains how they can open new directions in the logic combinational design and offer better automated solutions.

The major bottleneck of the evolutionary method is the S*calability of* E*volution* which refers to the fact that the evaluation time of a certain circuit grows exponentially with the increasing number of inputs. This means that it becomes quite difficult to evolve larger systems and in this case it is more efficient to split the large structure in smaller modular cells.

The genetic algorithms can be used in the automatic design of analogue and digital circuits as optimization methods. This could lead to better or competitive results with human designed cells. One successful method starts with a fixed circuit topology, chosen by the human designer's experience and the goal is to find the component values and transistors sizes in order to create a circuit having the desired functionality. There are many successful examples including filters and amplifiers but also different schemes of operational amplifiers at transistor level.

The use of genetic algorithm in analog circuit synthesis could increase with the help of progress in the analogue programmable integrated circuits.

Achieving reconfigurable adaptive systems in RF telecommunication applications with the help of genetic algorithms is a promising design method. It was proposed to be applied in tuning intermediate frequency filters for mobile telecommunications and also in designing programmable gain amplifiers.

Circuits' re-configurability is achieved with the help of capacitors or resistances networks which are capable to be switched in an established range. The components' values are digitally controlled using genetic algorithms which will react depending on the real values of the passive components which could be different in each designed chip.

The evolutionary methods could be very helpful in order to correct the variations of electronic components during the fabrication process and also to adapt their functionality to external unpredictable influences.

Biomedical engineering is another field where reconfigurable circuits are very useful. The auditory prosthesis electronic and other electronic simulators contain filters and variable gain amplifiers. The response of these circuits could be controlled by adaptive systems with the help of genetic algorithms. Especially when the stimulators are implanted, the adaptive response is necessary and it can be achieved by using evolvable hardware.

References

Abouellail, M., El-Araby, E., Taher, M., El-Ghazawi, T., Newby, G.: DNA and Protein Sequence Allingnment with High Performance Reconfigurable Systems. In: Arslan, T., Stoica, A., et al. (eds.) Proceedings of 2007 NASA/ESA Conference on Adaptive Hardware and Systems, Edinburgh, UK, August 5-8, pp. 334–341 (2007)

*** (Actel), Data book and design guide, Actel, Sunnyvale (1995)

Agnelli, F., et al.: System Analysis and Design of a Reconfigurable RF Front-End. IEEE Circuits and Systems Magazine, Special Issue on Wireless Reconfigurable Terminals (first Quarter, 2006)

Allen, P., Geiger, R.: VLSI Design Techniques for Analog and Digital Circuits. McGraw Hill, New York (1990)

Allen, P.E., Holberg, D.: CMOS Analog Circuit Design. Oxford University Press, Oxford (1987)

Ando, S., Iba, H.: Artificial Immune System for Classification of Gene Expression Data. In: Cantú-Paz, E., Foster, J.A., Deb, K., Davis, L., Roy, R., O'Reilly, U.-M., Beyer, H.-G., Kendall, G., Wilson, S.W., Harman, M., Wegener, J., Dasgupta, D., Potter, M.A., Schultz, A., Dowsland, K.A., Jonoska, N., Miller, J., Standish, R.K. (eds.) GECCO 2003, Part II. LNCS, vol. 2724, pp. 1926–1937. Springer, Heidelberg (2003)

Aoki, T., Homma, N., Higuchi, T.: Evolutionary Design of Arithmetic Circuits. IEICE Trans. Fundamentals E82-A(5), 798–806 (1999)

Arotaritei, D., Negoita, M.G.: Optimization of recurrent NN by GA with variable length genotype. In: McKay, B., Slaney, J.K. (eds.) Canadian AI 2002. LNCS, vol. 2557, pp. 681–692. Springer, Heidelberg (2002)

Attoh-Okine, N.: General Framework for Applying Artificial Immune Systems to a Highway Infrastructure Deterioration Modeling. In: Workshop Notes-Workshop 4: Immunity- Based Systems, International Conference on Multiagent Systems/International Symposium on Models for Software Architecture, Keihanna Plaza, Tokyo, Japan, pp. 25–37 (December 1996)

Azizi, N.: Automated Analog Circuit Design Using Genetic Algorithms, ECE1352 Reading Assignment, Department of Electrical and Computer Engineering University of Toronto (2002)

Baschirotto, A., et al.: Baseband Analog Front-End and Digital Back-End for Reconfigurable Multi-Standard Terminals. IEEE Circuits and Systems Mag. Special Issue on Wireless Reconfigurable Terminals (first Quarter, 2006)

Baez, F., Duster, J.S., Kornegay, K.T.: A low power 60 dB control range programmable gain amplifier in sige technology. In: IASTED Int. Conference 2004, pp. 477–481 (2004)

Baker, M.W., Sarpeshkar, R.: Low-Power Single-Loop and Dual-Loop AGCs for Bionic Ears. IEEE Journal of Solid-State Circuits 41(9) (September 2006)

Becker, J., Trendelenburg, S., Henrici, F., Manoli, Y.: A Field Programmable Gm-C Filter Aarray (FPAA) for Online Adaptation to Environmental Changes. In: NASA/ESA Conference on Adaptive Hardware and Systems 2007, Edinburgh, UK, pp. 547–553 (2007)

Bradley, D.W., Ortega-Sanchez, C., Tyrrell, A.: Embryonics + Immunotronics: A Bio-Inspired Approach to Fault Tolerance. In: Proceedings of the 2nd NASA/DoD Workshop on Evolvable Hardware, pp. 215–223. IEEE Computer Society, Los Alamitos (2000)

Bradley, D.W., Tyrrell, A.: The Architecture for a Hardware Immune System. In: Proceedings of the 3rd NASA/DoD Workshop on Evolvable Hardware, pp. 193–200. IEEE Computer Society, Los Alamitos (2001)

Bradley, D.W., Ortega-Sanchez, C., Tyrrell, A.: Embryonics + Immunotronics: A Bio- Inspired Approach to Fault Tolerance. In: Proceedings of the 2nd NASA/DoD Workshop on Evolvable Hardware, pp. 215–223. IEEE Computer Society, Los Alamitos (2000)

Brooks, R.: Intelligence without reason. In: Proceedings of IJCAI 1991, pp. 569–595 (1991); Canham, R., Jackson, H.A., Tyrrell, A.: Robot Error Detection Using an Artificial Immune System. In: Proceedings of the 3rd NASA/DoD Conference on Evolvable Hardware. IEEE Computer Society, Los Alamitos, pp. 199–206 (2003)

Bult, K., Geelen, G.J.G.M.: An inherently linear and compact MOST-only current division technique. IEEE J. Solid-State Circuits 27(12), 1730–1735 (1992)

Chung, W.-Y., Chuang, C.-C., Liao, Y.-L., Chang, C.-W.: A New Microstimulator with Pulse Width Modulation. Journal of Medical and Biological Engineering 24(3), 125–131 (2004)

Coello, C.A., Christiansen, A.D., Aguirre, A.H.: Using Genetic Algorithms to Design Combinational Logic Circuits. Intelligent Engineering through Artificial Neural Networks 6, 391–396 (1996)

Coello, C.A., Christiansen, A.D., Aguirre, A.H.: Automated Design of Combinational Logic Circuits using Genetic Algorithms. Intelligent Engineering through Artificial Neural Networks 6, 391–396 (1996)

Coello, C.A., Christiansen, A.D., Aguirre, A.H.: Automated design of combinational logic circuits using genetic algorithms. In: Proc. Int. Conf. on Artificial Neural Nets and Genetic Algorithms, pp. 335–338 (1997)

Csipkes, G., Hintea, S., Csipkes, D., Rus, C., Festila, L., Fernandez-Canque, H.: A Digitally Reconfigurable Low Pass Filter for Multi-mode Direct Conversion Receivers. In: Lovrek, I., Howlett, R.J., Jain, L.C. (eds.) KES 2008, Part III. LNCS (LNAI), vol. 5179, pp. 335–342. Springer, Heidelberg (2008)

Csipkes, D., Csipkes, G.: Synthesis method for state variable G_m-C filters with a reduced number of active components. In: MIXDES 2003, pp. 292–297 (2003)

Csipkes, G., Csipkes, D.: Synthesis method for G_m-C complex polyphase filter design. In: MIXDES 2003, Poland, pp. 286–291 (2003)

Csipkes, G., Csipkes, D.: A Programmable Channel Select Filter for Multi-Mode Wireless Applications. In: MIXDES 2004, Szcecin, Poland, June 24-26, pp. 145–150 (2004) ISBN 83-919289-7-7

Csipkes, G.: Integrated Realizations of Reconfigurable Low Pass and Band Pass Filters for Wide-Band Multi-Mode Receivers, PhD Thesis, TUD press (2006) ISBN 3-938863-46-3

Cutello, V., Nicosia, G., Pavone, M.: A Hybrid Immune Algorithm with Information Gain for the Graph Coloring Problem. In: Cantú-Paz, E., Foster, J.A., Deb, K., Davis, L., Roy, R., O'Reilly, U.-M., Beyer, H.-G., Kendall, G., Wilson, S.W., Harman, M., Wegener, J., Dasgupta, D., Potter, M.A., Schultz, A., Dowsland, K.A., Jonoska, N., Miller, J., Standish, R.K. (eds.) GECCO 2003, Part I. LNCS, vol. 2723, pp. 171–183. Springer, Heidelberg (2003)

Cutello, V., Nicosia, G.: Noisy Channel and Reaction-Diffusion Systems: Models for Artificial Immune Systems. In: Palade, V., Howlett, J.R., Jain, L. (eds.) Knowledge-Based Intelligent Information and Engineering Systems, Part II, pp. 496–502. Springer, Heidelberg (2003a)

Danforth, M., Levitt, K.: Immune System Model for Detecting Web Server Attacks. In: Arif Wani, M. (ed.) Proceedings of ICMLA 2003, LOS Angeles, CA, USA, June 23-24, pp. 161–167 (2003) ISBN 0-9741614-0-3

Dasgupta, D., Attoh-Okine, N.: Immunity-Based Systems: A Survey. In: Workshop Notes-Workshop 4: Immunity-Based Systems, International Conference on Multiagent Systems/International Symposium on Models for Software Architecture, Keihanna Plaza, Tokyo, Japan, pp. 13–25 (December 1996)

Dasgupta, D.: Artificial Neural Networks and Artificial Immune Systems: Similarities and Differences. In: Proceedings of The IEEE Systems, Man and Cybernetics Conference, vol. 1, pp. 873–878 (1997)

Deaton, R., Garzon, M.: Fuzzy logic with biomolecules. In: Soft Computing, vol. 5, pp. 2–9. Springer, Heidelberg (2001)

Deaton, R., Garzon, M., Rose, J.A., Murphy, R.C., Stevens Jr., S.E., Franceschetti, D.R.: A DNA Based Artificial Immune System for Self-Nonself Discrimination. In: Proceedings of The IEEE System, Man, and Cybernetics I, pp. 862–866 (1997)

De Castro, L.N., Von Zuben, F.J.: Immune and Neural Networks Models: Theoretical and Empirical Comparisons. International Journal of Computational Intelligence and Applications (IJCIA) 1(3), 239–259 (2001)

De Castro, L.N., Timmis, J.I.: Artificial Immune Systems: A New Computational Approach. Springer, London (2002)

Dellsperger, T., Burger, T., Maurer, L., Christen, T.: Reconfigurable RF Transceivers for Multi- Standard Terminals – An E2R View (2006)

Dilger, W.: The Immune System of the Smart Home. In: Workshop Notes-Workshop 4: Immunity-Based Systems, International Conference on Multiagent Systems/International Symposium on Models for Software Architecture, Keihanna Plaza, Tokyo, Japan, pp. 72–81 (December 1996)

Dong, M., Zhang, C., Mai, S., Wang, Z.. Li, D.: A wideband frequency-shift keying demodulator for wireless neural stimulation microsystems. Journal Zhejiang Univ. Science 7(6), 1056–1060 (2006)

Duong, Q.-H., Le, Q., Kim, C.-W., Lee, S.-G.: A 95-dB Linear Low-Power Variable Gain Amplifier. IEEE Trans. on Circuits and Systems 53(8), 1648–1657 (2006)

Fagarasan, F., Negoita, M.G.: A Genetic-Based Method for Learning the Parameter of a Fuzzy Inference System. In: Kasabov, N., Coghill, G. (eds.) Artificial Neural Networks and Expert Systems, pp. 223–226. IEEE Computer Society Press, Los Alamitos (1995)

Fang, W.-C., Sheu, B.J., Venus, H., Sandau, R.: Smart-Pixel Array Processor Based on Optimal Cellular Neural Networks for Space-Sensor Applications. In: Proc. IEEE ICCD, pp. 703–708 (1995)

Festila, L., Groza, R., Szolga, L., Hintea, S.: Log-Domain multipliers for VLSI architectures, Buletinul Stiintific al Universitatii Politehnica din Timisoara, Seria Electronica si Telecomunicatii, Tom 51(65), Fascicola 1-2 (2006)

Feştila, L., Groza, R., Cirlugea, M., Fazakas, A.: Modular analysis and design of log-domain circuits based on LINELIN transformations. Analog Integrated Circuits and Signal Processing 50(3), 231–249 (2006)

Festila, L., Neag, M.: A Unitary Analysis of Some Voltage- and Current-Mode Linear Circuits. Analog Integrated Circuits and Signal Processing, 5–24 (1999)

Feştila, L., Groza, R., Cirlugea, M., Fazakas, A.: Log-Domain Linear SVM Classifier, Mixed Design of Integrated Circuits and Systems. In: MIXDES 2007 Ciechocinek, June 21-23, pp. 301–304 (2007) ISBN: 83-922632-4-3

Festila, L., Topa, M., Hintea, S., Cirlugea, M., Groza, R.: A General Modular Design Of Elin Filters Based On F-1nf Models. In: IEEE-TTTC International Conference on Automation, Quality and Testing, Robotics, Cluj-Napoca, pp. 227–232 (2004) ISBN 973-713-047-2

Festila, et al.: Support Vector Machines, Editura Casa Cărţii, Cluj -Napoca (2007) ISBN 978-973-133-107-2

Forrest, S., Perelson, A.: Genetic Algorithms and the Immune Systems. In: Goos, C., Hartmanis, J. (eds.) Proceedings of The Parallel Problem Solving from Nature, pp. 320–325. Springer, Heidelberg (1990)

Forrest, S., Javornik, B., Smith, R.E., Perelson, A.: Using Genetic Algorithms to Explore Pattern Recognition in the Immune System. Evolutionary Computation 1(3), 191–211 (1993)

Furuhashi, T., Nakaoka, K., Uchikawa, Y.: A New Approach to Genetic Based Machine Learning and an Efficient Finding of Fuzzy Rules. In: Proceedings of the IEEE/Nagoya University WWW on Fuzzy Logic and Neural Networks/Genetic Algorithms, Nagoya, Japan, pp. 114–122 (1994)

Gajda, Z., Sekanina, L.: Reducing the Number of Transistors in Digital Circuits Using Gate-Level Evolutionary Design. In: GECCO 2007, London (2007)

Galbraith, D., Soma, M., White, R.: A wide-band efficient inductive transdermal power and data link with coupling insensitive gain. IEEE Trans. Biomed. Eng. 34, 265–275 (1987)

Garrison, W.G., Ramsden, E., Ahmed, S.: An Empirical Comparison of Evolutionary Algorithms for Evolvable Hardware with Minimum Time-to-Reconfigure Requirements. In: Proceedings of 2003 NASA DoD Conference on Evolvable Hardware, Chicago Illinois, USA, pp. 59–66 (July 2003) ISBN 0-7695-1977-6

Garzon, M.: Biomolecular Machines and Artificial Evolution. Genetic Programming and Evolvable Machines 4(2), 107–111 (2003)

Garzon, M., Blain, D., Bobba, K., Neel, A., West, M.: Self-Assembly of DNA-like Structures In Silico. Genetic Programming and Evolvable Machines 4(2), 185–200 (2003)

Gata, D.G., et al.: A 1,1-V 270- μ A Mixed-Signal Hearing Aid Chip. IEEE Journal of Solid-State Circuits 37(12) (2002)

Gehani, A., Reif, J.: Micro Flow Bio-Molecular Computation. In: Proceedings of the fourth DIMACS Worskhop on DNA-based Computers, PENSYLVANIA, USA, pp. 253–266 (1998)

Germanovix, W., Toumazou, C.: Design of a Micropower Current-Mode Log-Domain Analog Cochlear Implant. IEEE Transactions on Circuits and Systems-II: Analog and Digital Signal Processing 47(10) (October 2000)

Gers, F., de Garis, H.: CAM-Brain: A New Model for ATR's Cellular Automata Based Artificial Brain Project. In: Higuchi, T., Iwata, M., Liu, W. (eds.) Evolvable Systems: From Biology to Hardware, pp. 437–455. Springer, Heidelberg (1997)

Ghovanloo, M.: A wideband frequency-shift keying wireless link for inductively powered biomedical implants. IEEE Trans. Circuits & Systems – 1 51(12), 2374–2383 (2004)

Goonatilake, S., Treleaven, P.: Intelligent Systems for Finance and Business. John Wiley & Sons, Chichester (1996)

Gordon, G.W., Bentley, P.J.: On Evolvable Hardware. In: Ovaska, S., Sztandera, L. (eds.) Soft Computing in Industrial Electronics (2002)

Groza, R., Festila, L., Hintea, S., Cirlugea, M.: Log-Domain Binary SVM Image Classifier. In: Lovrek, I., Howlett, R.J., Jain, L.C. (eds.) KES 2008, Part III. LNCS (LNAI), vol. 5179, pp. 368–375. Springer, Heidelberg (2008)

Hammerschmied, C.M.: CMOS A/D Converters Using MOSFET-Only R-2R Ladders, PhD Thesis, Zurich (2000)

Harding, S., Miller, J.: Evolution in materio: Initial Experiments with Liquid Crystal. In: Proceedings of the 6th NASA/DoD Conference on Evolvable Hardware, pp. 167–176. IEEE Computer Society Press, Los Alamitos (2004)

Harrison, R.R., Watkins, P.T., Kier, R.J., Lovejoy, R.O., Black, D.J., Greger, B., Solzbacher, F.: A Low-Power Integrated Circuit for a Wireless 100-Electrode Neural Recording System. IEEE Journal of Solid-State Circuits 42(1), 123–133 (2007)

Higuchi, T., Iwata, M., Kajtani, I., Iba, H., Hirao, Y., Furuya, T., Manderick, B.: Evolvable Hardware and Its Applications to Pattern Recognition and Fault-Tolerant Systems. In: Sanchez, E., Tomassini, M. (eds.) Towards Evolvable Hardware, pp. 118–136. Springer, Heidelberg (1996)

Higuchi, T.: Evolvable Hardware with Genetic Learning. In: Proceedings of IEEE International Symposium on Circuits and Systems, ISCAS 1996, Atlanta, USA, May 13 (1996)

Higuchi, T., Iba, H., Manderick, B.: Evolvable Hardware. In: Kitano, H., Hendler (eds.) Massively Parallel Artificial Intelligence, pp. 339–420. AAAI Press/ MIT Press (1994)

Hintea, S., Csipkes, G., Rus, C., Csipkes, D., Fernandez-Canque, H.: On the Design of a Reconfigurable OTA-C Filter for Software Radio. In: Second NASA/ESA Conference on Adaptive Hardware and Systems. AHS 2007, August 5-8, pp. 541–546 (2007)

Hintea, S., Simion, E., Feştilă, L.: Radiofrequency link used in partially-implanted auditory prosthesis. In: Proc. of the 3rd IEEE Int.Conference on Electronics, Circuits and Systems. ICECS 1996, Rhodos, pp. 1143–1146 (1996)

Hintea, S., Mihu, I.: Class DE Amplifiers and their medical applications. In: Proc. of The 6th International Conference on Optimization of Electrical and Electronic Equipments, OPTIM 1998, Brasov, May 14-15 (1998)

Hintea, S.: A New Method to measure impedance without contact. ACTA TEHNICA NAPOCENSIS, Electronics and Telecommunications, Nr.1/1998, Cluj-Napoca (1998)

Hintea, S., Fazakas, A., Cârlugea, M., Petrescu, D.: Developing a control system for an elevator using XC 4000XL family of Xilinx FPGA. Acta Tehnica Napocensis Electronics and Telecommunications, No.1/2004, Cluj-Napoca (2004)

Hintea, S., Feştilă, L., Lupea, D., Chilian, A.: Analysis and synthesis of class D amplifiers used in auditory prosthesis. In: Proceedings of MIXDES 1996, Lodz, pp. 199–204 (1996)

Hoffman, G.W.: A Neural Network Model Based on the Analogy with the Immune System. Journal of Theoretical Biology 122, 33–67 (1986)

Hueber, G., Maurer, L., et al.: On the Design of a Multi-Mode Receive Digital – Front – End for cellular terminal RFICs

Ishida, Y.: The Immune System as a Self-Identification Process: a Survey and a Proposal. In: Workshop Notes-Workshop 4: Immunity-Based Systems, International Conference on Multiagent Systems/International Symposium on Models for Software Architecture, Keihanna Plaza, Tokyo, Japan, pp. 2–13 (December 1996)

Ishida, Y.: Agent-Based Architecture of Selection Principle in the Immune System. In: Workshop Notes-Workshop 4: Immunity- Based Systems, International Conference on Multiagent Systems/International Symposium on Models for Software Architecture, Keihanna Plaza, Tokyo, Japan, pp. 93–105 (December 1996b)

Ishida, Y., Adachi, N.: An Immune Algorithm for Multiagent: Application to Adaptive Noise Neutralization. In: IEEE/RSJ Int. Conf. On Intelligent Robots and Systems (IROS 1996), pp.1739–1746 (1996)

Ishida, Y.: Immunity-Based Systems: A Design Perspective. Springer, Heidelberg (2004)

Ishiguro, A., Watanabe, Y., Kondo, T., Shirai, Y., Uchikawa, Y.: Immunoid: A Robot with a Decentralized Behavior Arbitration Mechanisms Based on the Immune System. In: Workshop Notes-Workshop 4: Immunity-Based Systems, International Conference on Multiagent Systems/International Symposium on Models for Software Architecture, Keihanna Plaza, Tokyo, Japan, pp. 82–92 (December 1996)

Jang, R.J.-S.: ANFIS, Adaptive-Network-Based Fuzzy Inference System. IEEE Transaction on System, Man, and Cybernetics 23(3), 665–685 (1993)

Jeon, J.I., Lee, H., Rhee, K.P., Lee, C.: Knowledge-Based Intelligent Information & Engineering Systems. In: Khosla, R., Howlett, J.R., Jain, L. (eds.) Proceedings of KES 2005, Melbourne, Australia (2005)

Jerne, N.K.: The Immune System. Scientific American 229(1), 52–60 (1973)

Jerne, N.K.: The Generastive Grammar of the Immune System. The EMBO Journal 4(4), 847–852 (1985)

Jenkins, H.J.: Designing with FPGAs and CPLDs. Prentice-Hall, Englewood Cliffs (1994)

Jun, J.-H., Lee, D.-W., Sim, K.-B.: Realization of Cooperative Behavior in Distributed Autonomous Robotic Systems Using Artificial Immune System. In: Proceedings of the IEEE of The IEEE System, Man, and Cybernetics, vol. 4, pp. 614–619 (1999)

Kalganova, T.: Bidirectional Incremental Evolution in Extrinsic Evolvable Hardware. In: Proc. of the 2nd NASA/DoD Workshop on Evolvable Hardware, pp. 65–74. IEEE Computer Society, Los Alamitos (2000)

Karnaugh, M.: A map method for synthesis of combinational logic circuits. AIEE Trans. Communications and Electronic 72(I), 593–599 (1953)

Kajitani, I., Hoshino, T., Nishikawa, D., Yokoi, H., Nakaya, S., Yamauchi, T., Inuo, T., Kajihara, N., Iwata, M., Keymeulen, D., Higuchi, T.: A Gate-level EHW Chip: Implementing GA Operations and Reconfigurable Hardware on a Single LSI. In: Sipper, M., Mange, D., Pérez-Uribe, A. (eds.) ICES 1998. LNCS, vol. 1478, p. 1. Springer, Heidelberg (1998)

Kelsey, J., Timmis, J.I.: Immune Inspired Somatic Contiguous Hypermutation for Function Optimization. In: Cantú-Paz, E., Foster, J.A., Deb, K., Davis, L., Roy, R., O'Reilly, U.-M., Beyer, H.-G., Kendall, G., Wilson, S.W., Harman, M., Wegener, J., Dasgupta, D., Potter, M.A., Schultz, A., Dowsland, K.A., Jonoska, N., Miller, J., Standish, R.K. (eds.) GECCO 2003. LNCS, vol. 2723, pp. 207–218. Springer, Heidelberg (2003)

Kelly, M.: Fit for Future? Evolutionary Computing in Industry, EvoNews, Issue 2, pp. 1–3 (December 1996)

Kenington, P.B.: RF and Baseband Techniques for Software Designed Radio. Artech House, Boston (2005)

Khosla, R., Dillon, T.: Engineering Intelligent Hybrid Multi-Agent Systems. Kluwer Academic Publishers, Dordrecht (1997)

Kim, T.W., Kim, B., Cho, Y., Kim, S., Kim, B., Lee, K.: Low Power 60 dB Gain Range with 0.25 dB Resolution CMOS RF Programmable Gain Amplifier for Dual-Band DAB/T-, DMB Tuner IC. IEEE Trans. on Circuits and Systems, 133–136 (2005)

Ko, W.H., Liang, S.P., Fung, C.D.: Design of radio- frequency powered coils for implant instruments. Med. Bio. Eng. Comput. 15, 634–640 (1977)

Kountouris, A., Moy, C., Rambaud, L.: Reconfigurability: A Key Property in Software Radio Systems. In: First Karlshruhe Workshop on Software Radios, March 29-30, Karlshruhe, Allemagne (2000)

Lee, M.A., Takagi, H.: Integrating Design Stages of Fuzzy Systems Using Genetic Algorithms. In: Proc IEEE Int Conf on Fuzzy Systems (FUZZ-IEEEE 1993), San Francisco, California, pp. 612–617 (1993)

Leiviska, K. (ed.): eunite Roadmap, Smart Adaptive Systems, state-of-the-art and challenging new application and research areas (2004), http://www.eunite.org

Loh, K.H., Hiser, D.L., Adams, W.J., Geiger, R.L.: A Versatile Digitally Controlled Continuous-Time Filter Structure with Wide Range and Fine Reolution Capability. IEEE Trans on Circuits and Systems-II. Analog and Digital Signal Processing 32(5), 265–276 (1992)

Lohn, J.D., Colombano, S.P.: A Circuit Representation Technique for Automated Circuit Design. IEEE Trans. on Evolutionary Computation 3, 205–219 (1999)

Lohn, J.D., Haith, G.L., Colombana, S.P., Stassinopoulos, D.: Towards Evolving Electronic Circuits for Autonomous Space Applications. In: Proceedings of the 2000 IEEE Aerospace Conference, vol. 5(1), pp. 476–486 (2000)

Loizou, P.C.: Mimicking the Human Hear. IEEE Signal Processing Magazine, 101–130 (1998)

Louis, S.J.: Genetic Algorithms as a Computational Tool for Design, Ph.D. thesis, Department of Computer Science, Indiana University (August 1993)

Lupea, D., Pursche, U., Jentschel, H.-J.: Spectral Signature Analysis – Bist for RF Front – Ends, Advances in Radio Science, Copernicus GmbH, pp. 155–160 (2003)

Mak, P.M., Seng-Pan, U., Martins, R.P.: On the Architectural Design of Frequency Agile Multi-standard Wireless Receivers. IEEE Circuits and Systems Magazine, 6–25 (Second Quarter, 2007)

Marchal, P., Nussbaum, P., Piguet, C., Durand, S., Mange, D., Sanchez, E., Stauffer, A., Tempesti, G.: Embryonics: The Birth of Synthetic Life. In: Sanchez, E., Tomassini, M. (eds.) Towards Evolvable Hardware, pp. 166–195. Springer, Heidelberg (1996)

Marchal, P., Piguet, C., Mange, D., Stauffer, A., Durand, S.: Embryological Development on Silicon. In: Brooks, A.R., Maes, P. (eds.) Artificial Life IV, pp. 365–370. The MIT Press, Cambridge (1994)

Maurer, L., et al.: On the Design of a Configurable UMTS/NAVSAT Transceiver. In: 14th IST Mobile & Wireless Communications Summit, Dresden (June 2005)

Maurer, L., et al.: On the Architectural Design of Frequency – Agile Multi-Standard Wireless Receivers. In: 14th IST Mobile & Wireless Communications Summit, Dresden (June 2005)

Melgarejo, M., Pena-Reyes, A.C.: Implementing Interval Type-2 Fuzzy Processors. IEEE COMPUTATIONAL Intelligence Magazine 2(1), 63–71 (2007), http://www.ieee.cis.org

Melgarejo, M., Garcia, A., Pena-Reyes, A.C.: Computational Model and Architectural Proposal for a Hardware Type-2 Fuzzy System. In: Proceedings of IASTED Conference on Neural Networks and Computational Intelligence, Grindenwald, Switzerland, pp. 279–284 (2004)

Mellis, M.: Evolutionary Computing at Space Engineering S.p.A. Rome, EvoNews (2), 3–6 (December 1996)

Mihaila, D., Fagarasan, F., Negoita, M.G.: Architectural Implications of Genetic Algorithms Complexity in Evolvable Hardware Implementation. In: Proceedings of the European Congress EUFIT 1996, Aachen, Germany, vol. 1, pp. 400–404 (September 1996)

Miller, J.F., Downing, K.: Evolution in materio: Looking Beyond the Silicon Box. In: Stoica, A., Lohn, J., et al. (eds.) Proceedings of 2002 NASA DoD Conference on Evolvable Hardware, Alexandria Virginia, USA, pp. 167–176 (July 2002) ISBN 0-7695-1718-8

Miller, J.F., Thomson, P., Fogarty, T.C.: Designing Electronic Circuits Using Evolutionary Algorithms. Arithmetic Circuits: A Case Study. In: Quagliarella, D., Periaux, J., Poloni, C., Winter, G. (eds.) Genetic Algorithms and Evolution Strategies in Engineering and Computer Science: Recent Advancements and Industrial Applications, pp. 103–131. Wiley, Chichester (1997)

Miller, J.F., Thompson, A., Thompson, P., Fogarty, T.C. (eds.): ICES 2000. LNCS, vol. 1801. Springer, Heidelberg (2000)

Miller, J., Thomson, P.: Aspects of digital evolution: Geometry and learning. In: Sipper, M., Mange, D., Pérez-Uribe, A. (eds.) ICES 1998. LNCS, vol. 1478, pp. 25–35. Springer, Heidelberg (1998)

Mills Jr., A.P., Turberfield, M., Turberfield, J.A., Yurke, B., Platzman, P.M.: Experimental aspects of DNA neural network computation. In: Soft Computing, vol. 5, pp. 10–18. Springer, Heidelberg (2001)

Mitola, J.: The Software Radio Architecture. IEEE Communications Magazine, 26–38 (1995)

McCluskey, E.J.: Minimization of Boolean function. Bell Systems Technical Journal 35(5), 1417–1444 (1956)

Murakawa, M., Yoshizawa, S., Adachi, T., Suzuki, S., Takasuka, K., Iwata, M., Higuchi, T.: Analogue EHW chip for intermediate frequency filters. In: Sipper, M., Mange, D., Pérez-Uribe, A. (eds.) ICES 1998. LNCS, vol. 1478, pp. 134–143. Springer, Heidelberg (1998)

Murakawa, M., Adachi, T., Niino, Y., Kasai, Y., Takahashi, E., Takasuka, K., Higuchi, T.: An AI-calibrated IF filter: a yield enhancement method with area and power dissipation reductions. IEEE Journal of Solid-State Circuits 38(3), 495–502 (2003)

Negoita, M.G., Sekanina, L.: Adaptive and Evolvable Hardware and Systems: the State of the Art and the Prospectus for Future Developments. KES – The International Journal of Knowledge-based Intelligent Engineering Systems (in press) (2007) ISSN 1327-2314

Negoita, M.G.: Artificial Immune Systems – An Emergent Technology Performing the Defense of Complex Systems. In: Proceedings of BLISS 2007 (2007 ECSIS Symposium on Bio-inspired, Learning and Intelligent Systems for Security), Edinburgh, Scotland, UK, pp. 145–148 (2007) ISBN 0-7695-2919-4

Negoita, M.G., Neagu, D., Palade, V.: Computational Intelligence: Engineering of Hybrid Systems. Springer, Heidelberg (2005)

Negoita, M.G.: Artificial immune systems—an emergent technology for autonomous intelligent systems and data mining. In: Gorodetsky, V., Liu, J., Skormin, V.A. (eds.) AIS-ADM 2005. LNCS, vol. 3505, pp. 18–36. Springer, Heidelberg (2005a)

Negoita, M.G.: Basics of Engineering the Hybrid Intelligent Systems - Not only Industrial Applications. In: Negoita, M.G., Reusch, B. (eds.) Real World Applications of Computational Intelligence, pp. 5–55. Springer, Heidelberg (2005)

Negoita, M.G., Pritchard, D.: Using Virtual Student Model for Testing Intelligent Tutoring Systems. J. Interactive Technology & Smart Education 1, 3–10 (2004)

Negoita, M.G., Stoica, A.: Evolvable Hardware in the Frame of Computational Intelligence – Implication on Engineering Design and Intelligence of Autonomous Systems. In: A tutorial course at KBCS 2004, Hyderabad, India (2004)

Negoita, M.G., Pritchard, D.: Testing Intelligent Tutoring Systems by Virtual Students. In: Arif Wani, M. (ed.) Proc. Int. Conf. on Machine-Learning and Applications (ICMLA 2003), Los Angeles, USA, pp. 98–104 (2003)

Negoita, M.G.: A Modern Solution for the Technology of Preventing and Alarm Systems: Evolutionary Computation in Evolvable Hardware Implementation. In: Proceedings of The Second Dunav Preving International Conference on Preventing and Alarm Systems, Belgrade, pp. 201–209 (November 1997)

Negoita, M.G., Dediu, A.H., Mihaila, D.: Application of GA with Local Improvement of the Chromosomes for the Design of EHW Architectures. In: Proceedings of the European Congress EUFIT 1997, Aachen, Germany, pp. 814–818 (September 1997)

Negoita, M.G.: Methods Based on Evolutionary Computation Techniques for Implementation of Evolvable Hardware. In: Proceedings of the International Conference of Technical Informatics, Timisoara, Romania. Computer Science and Engineering, vol. 1, pp. 37–45 (November 1996)

Negoita, M.G.: Evolutionary Computation in the Soft Computing Framework. In: Zimmermann, H.-J., Negoita, M.G., Dascalu, D. (eds.) Real World Applications of Intelligent Technologies, Editura Academiei Romane, Bucharest, pp. 113–139 (1995)

Negoita, M.G., Z.H.-J.: Genetic Algorithms in the Soft Computing Framework, Application in Complex Cybernetic Systems, Cybernetica XXXVIII(1) (1995)

Nikolaev, N.I., Iba, H., Slavov, V.: Inductive Genetic Programming with Immune Network Dynamics. Advances in Genetic Programming (3), 355–376, 59–68 (1999)

Oda, T., White, T.: Developing an Immunity to Spam. In: Cantu-Paz, E., et al. (eds.) GECCO 2003. LNCS, vol. 2723, pp. 207–242. Springer, Heidelberg (2003)

Okamoto, T., Watanabe, T., Ishida, Y.: Towards an Immunity-Based System for Detecting Masqueraders. In: Palade, V., Howlett, J.R., Jain, L. (eds.) Knowledge-Based Intelligent Information and Engineering Systems, Part II, pp. 488–495. Springer, Heidelberg (2003)

Okamoto, T., Watanabe, T., Ishida, Y.: Mechanism for Generating Immunity-Based Agents that Detect Masqeraders. In: Negoita, M., Howlett, J.R., Jain, L. (eds.) Knowledge-Based Intelligent Information and Engineering Systems, Part II, pp. 534–540. Springer, Heidelberg (2004)

Oltean, G., Hintea, S., Sipos, E.: Computational Intelligence in Analog Circuits Design. In: Lovrek, I., Howlett, R.J., Jain, L.C. (eds.) KES 2008, Part III. LNCS (LNAI), vol. 5179, pp. 384–391. Springer, Heidelberg (2008)

Oltean, G.: Fuzzy Techniques in Analog Circuit Design. WSEAS Transactions on Circuits and Systems 7(5), 402–415 (2008)

Oltean, G.: FADO - A CAD Tool for Analog Modules. In: Proc. of the International Conference on Computer as a Tool, EUROCON 2005, Belgrade, pp. 515–518 (2005) ISBN 1-4244-0050-3

Oltean, G., Miron, C., Mocean, E.: Multiobjective Optimization Method for Analog Circuits Design based on Fuzzy Logic. In: Proceedings of the 9th IEEE International Conference on Electronics, Circuits and Systems - ICECS 2002, Dubrovnik, Croatia, September 15-18, pp. 777–780 (2002) ISBN 0-7803-7597-1

Oltean, G., Hintea, S., Lupea, D.: A Fuzzy Optimization Engine for Analog Circuit Design. In: Proc. of the 8th International Conference MIXDES 2001, Zakopane, Poland, June 21-23, pp. 77–82 (2001) ISBN 83-87202-98-3

Oltean, M.: Switchable Glass: A possible medium for Evolvable Hardware. In: Stoica, A., Arslan, T., et al. (eds.) Proceedings of 2006 NASA/ESA Conference on Adaptive Hardware and Systems, Istanbul, Turkey, June 15-18, pp. 81–87 (2006) ISBN0-7695-2614-4

Ortega, C., Mange, D., Smith, S., Tyrrell, A.: Embryonics: A Bio-Inspired Cellular Architecture with Fault-Tolerant Properties. J. Genetic Programming and Evolvable Machines 1-3, 187–215 (2000)

Palade, V., Bumabaru, S., Negoita, M.: A Method for Compiling Neural Networks into Fuzzy Rules Using Genetic Algorithms and Hierarchical Approach. In: Proc. KES Int. Conf. on Knowledge-Based Intelligent Electronic Systems, Adelaide, Australia, pp. 353–358 (1998)

Pankiewicz, B., et al.: A Field Programmable Analog Array for CMOS Continuous-Time Filter Applications. IEEE JSSC 37(2) (2002)

Paun, G., Rozenberg, G., Salomaa, A.: DNA Computing – New Computing Paradigms. Springer, Heidelberg (1998)

Pavan, S., Tsividis, Y., Nagaraj, K.: A 60-350 MHz Programmable Analog Filter in a Digital CMOS Process. In: Proceedings of the European Solid State Circuits Conference, Duisburg, Germany, September 21-23 (1999)

Pavan, S., Tsividis, Y., Nagaraj, K.: Widely Programmable High Frequency Continuous Time Filters in Digital CMOS Technology. IEEE Journal of Solid State Circuits 35(4) (April 2000)

Pavan, S., Tsividis, Y.: Time Scaled Electrical Networks - Properties and Applications in the Design of Programmable Analog Filters. IEEE Transactions on Circuits and Systems - Analog and Digital Signal Processing 47(2), 161–165 (2000)

Pecenka, T., Kotasek, Z., Sekanina, L., Strnadel, J.: Automatic discovery of rtl benchmark circuits with predefined testability properties. In: Stoica, A., Arslan, T., et al. (eds.) Proceedings of 2005 NASA/DoD Conference on Evolvable Hardware, pp. 51–58. IEEE Computer Society, Los Alamitos (2005)

Pena, J., Upegui, A.: A Population-oriented Architecture for Particle Swarms. In: Arslan, T., Stoica, A., et al. (eds.) Proceedings of 2007 NASA/ESA Conference on Adaptive Hardware and Systems, Edinburgh, UK, August 5-8, pp. 563–570 (2007)

Perciante, A.: The Approach of Selex Communications to Software Defined Radio. In: SDR International Conference WSR 2006, Karlsruhe (March 2006)

Percus, J.K., Percus, O., Person, A.S.: Predicting the Size of the Antibody Combining Region from Consideration of Efficient Self/Nonself Discrimination. Proceedings of The National Academy of Science 60, 1691–1695 (1993)

Pohlheim, H.: GEATbx Introduction to Evolutionary Algorithms: Overview, Methods and Operators, version 3.7 (November 2005), http://www.geatbx.com/

Pun, K.P., Choy, C.S., Chan, C.F., de Franca, J.E.: Current-Division Based Digital Frequency Tuning for Active RC Filters. Analog Integrated Circuits and Signal Processing, 61–69 (2005)

Quine, W.V.: A way to simplify truth function. American Mathematical Monthly 62(9), 627–631 (1955)

Rabaey, J.M., Chandrakasan, A., Nikolic, B.: Digital Integrated Circuits. A design perspective. Prentice Hall, Englewood Cliffs (2003)

Razavi, B.: RF Microelectronics. Prentice Hall PTR, Englewood Cliffs (1998)

Reis, C., Tenreiro Machado, J.A., Boaventura, C.J.: Evolutionary Design of Combinational Logic Circuits. Journal of Advanced Computational Intelligence and Intelligent Informatics 8(5), 507–513 (2004)

Ren, X., Kazmierski, T.: Performance Modeling and Optimization of RF Circuits Using Support Vector Machines. In: Proc. of 14th International Conference on Mixed Design of Integrated Circuits and Systems, MIXDES 2007, pp. 317–321 (2007)

Robinson, G.: Load Cell Shape Optimization Using Genetic Algorithm. EvoNews (2), 12–14 (December 1996)

Rose, A.J., Deaton, R., Franceschetti, D.R., Garzon, M., Stevens Jr., E.S.: A Statistical Mechanical Treatment of Error in the Annealing Biostep of DNA Computation. In: Proceedings of Special Program in DNA and Molecular Computing at GECCO 1999, Orlando, Florida, pp. 1829–1883. Morgan Kaufman, San Francisco (1999)

Rus, C., Festila, L., Csipkes, G., Hintea, S., Csipkes, D.: A Reconfigurable and Programmable Filter for Software Defined Radio Based on a Transconductor-Capacitor Analog Array, Cairo (2007)

Rus, C.M., Festila, L., Hintea, S.: A Programmable Band Pass Filter for Software Defined Radio Wireless Transceivers. In: IEEE SiPS Conference, Greece, Athens, November 2-4 (2005)

Rus, C.M., Hintea, S.: Corner Frequency Programming Strategy for Low Pass Filters in Software Defined Radio. Acta Technica Napocensis (ATN) (January 2005)

Rus, C.M., Festila, L., Hintea, S.: Linearity Considerations for Adaptively Biased Transconductors with Applications in Continuous Time Filters. Buletinul Stiintific al Universitatii Politehnica din Timisoara, Seria Electronica si Telecomunicatii, Tom 51(65), Fascicola, 1–2 (2006)

Rus, C.M., Festila, L., Hintea, S.: Programmable Filter Using Transconductor Elements for Reconfigurable Wireless Receivers. In: OPTIM 2006, Brasov, Romania, May 18-22 (2006)

Rus, C.M., Festila, L., Csipkes, G., Hintea, S., Csipkes, D.: Reconfigurable and Programmable Filter for Software Defined Radio. In: Proceedings of IEEE International Design and Test Workshop IDT 2007, Cairo, pp. 27–32 (december 2007)

Sanchez, E., Perez-Uribe, A., Upegui, A., Thoma, Y.: PERPLEXUS: Pervasive Computing Framework for Modeling Complex Virtually-Unbounded Systems. In: Arslan, T., Stoica, A., et al. (eds.) Proceedings of 2007 NASA/ESA Conference on Adaptive Hardware and Systems, Edinburgh, UK, August 5-8, pp. 587–592 (2007)

Sacristán, J., Marin, D., Oses, M.T.: Telemetric Implantable System for Neural Stimulation. In: IFESS 2002, 7th Annual Conference, Ljubljana, Slovenia (June 2002)

Sanchez, E., Mange, D., Sipper, M., Tomassini, M., Perez-Uribe, A., Stauffer, A.: Phylogeny, Ontogeny, and Epigenesis: Three Sources of Biological Inspiration for Softening Hardware. In: Higuchi, T., Iwata, M., Liu, W. (eds.) Evolvable Systems: From Biology to Hardware, pp. 35–55. Springer, Heidelberg (1997)

Sanchez, E.: Field Programmable Gate Arrays (FPGA) Circuits. In: Sanchez, E., Tomassini, M. (eds.) Towards Evolvable Hardware, pp. 1–18. Springer, Heidelberg (1996)

Santini, C.C., Zebulum, R.S., Pacheco, M.A.C., Velasco, M.M.R., Szwarcman, M.H.: PAMA – Programmable Analog Multiplexer Array. In: Proceedings of The 2001 NASA/DoD Conference on Evolvable Hardware, Long Beach, California, July 12-14, pp. 36–43 (2001)

Schelmbauer, W., Pretl, H., Maurer, L., Weigel, R., Adler, B., Fenk, J.: A Fully Integrated Analog Baseband IC for an UMTS Zero - IF Receiver (2000)

Sekanina, L.: Evolvable Components: From Theory to Hard ware Implementations. Natural Computing Series. Springer, Heidelberg (2004)

Sekanina, L.: Evolutionary Design of Digital Circuits: Where Are Current Limits? In: Stoica, A., Arslan, T., et al. (eds.) Proceedings of 2006 NASA/ESA Conference on Adaptive Hardware and Systems, Istanbul, Turkey, June 15-18, pp. 171–178 (2006)

Serdijn, W.A., Woerd, A.C., Davidse, J., Roermund, A.H.M.: A Low-Voltage Low-Power Fully-Integrable Automatic Gain Control for Hearing Instruments. IEEE Journal of Solid-State Circuits 29(8) (August 1994)

Shacklefrod, B., Okushi, E., Yasuda, M., Koizumi, H., Seo, K., Iwamoto, T., Yasuura, H.: High-performance hardware design and implementation of genetic algorithms. In: Teodorescu, H.-N., Jain, L.C., Kandel, A. (eds.) Hardware Implementation of Intelligent Systems, pp. 53–91. Springer, Heidelberg (2001)

Shimonara, K.: Evolutionary Systems for Brain Communications – Towards an Artificial Intelligence. In: Brooks, A.R., Maes, P. (eds.) Artificial Life IV, pp. 4–7. MIT Press, Cambridge (1994)

Shimooka, T., Shimizu, K.: Idiotypic Network Model for Feature Extraction in Pattern Recognition. In: Palade, V., Howlett, J.R., Jain, L. (eds.) Knowledge-Based Intelligent Information and Engineering Systems, Part II, pp. 511–518. Springer, Heidelberg (2003)

Shimooka, T., Shimizu, K.: Artificial Immune System for Personal Identification with Finger Vein Pattern. In: Negoita, M., Howlett, J.R., Jain, L. (eds.) Knowledge-Based Intelligent Information and Engineering Systems, Part II, pp. 511–518. Springer, Heidelberg (2004)

Silva, A., Guilherme, J., Neves, R.F., Horta, N.: Designing Reconfigurable Multi-Standard Analog Baseband Front-End for. 4G Mobile Terminals: System Level Design. In: Proc. Conf. on Telecommunications - ConfTele, Peniche, Portugal, vol. 1, pp. 117–120 (May 2007)

Slowik, A., Bialko, M.: Evolutionary design and optimization of combinational digital circuits with respect to transistor count. Bulletin of the Polish Academy of Sciences, Technical Sciences 54(4), 437–442 (2006)

Somani, A., Chakrabarti, P.P., Patra, A.: An Evolutionary Algorithm-Based Approach to Automated Design of Analog and RF Circuits Using Adaptive Normalized Cost Functions. IEEE Transactions on Evolutionary Computation 11(3), 336–353 (2007)

Springer, A., Weigel, R.: RF microelectronics for W-CDMA mobile communication systems. Electronics & Communication Engineering Journal 14(3), 92–100 (2002)

Stehr, U., Henkel, F., Dalluege, L., Waldow, P.: A Fully Differential CMOS Integrated 4th Order Reconfigurable Gm-C Low Pass Filter for Mobile communications. In: ICECS 2003, vol. 1, pp. 144–147 (2003)

Steyaert, M., Crols, J.: CMOS wireless transceiver design. Kluwer, Dordrecht (1997)

Stieglitz, T., Schuettler, M., Koch, K.P.: Implantable Biomedical Microsystems for Neural Prostheses. IEEE Engineering in Medicine and Biology Magazine, 58–66 (September-October 2005)

Stoica, A., Andrei, R.: Adaptive and Evolvable Hardware – A Multifaceted Analysis. In: Arslan, T., Stoica, A., et al. (eds.) Proceedings of 2007 NASA/ESA Conference on Adaptive Hardware and Systems, Edinburgh, UK, August 5-8, pp. 486–496 (2007)

Stoica, A., Zebulum, R.S., Ferguson, M.I., Keymeulen, D., Duong, V., Guo, X.: Evolving Circuits in Seconds: Experiments with a Stand-Alone Board-Level Evolvable System. In: Proceedings of The 2002 NASA/DoD Conference on Evolvable Hardware, Alexandria, Virginia, July 15-18, pp. 67–74 (2002)

Stoica, A., Zebulum, R., Keymeulen, D., Tawel, R., Daud, T., Thakoor, A.: Reconfigurable VLSI Architectures for Evolvable Hardware: From Experimental Field Programmable Transistor Arrays to Evolution-Oriented Chips. IEEE Trans. on Very Large Scale Integration (VLSI) Systems 9(1) (February 2001)

Stoica, A., Keymeulen, D., Zebulum, R., Thakoor, A., Daud, T., Klimeck, G., Jin, Y., Tawel, R., Duong, V.: Evolution of Analog Circuits on Field Programmable Transistor Arrays. In: Proceedings of the 2nd NASA/DoD workshop on Evolvable Hardware, July 13-15, p. 99 (2000)

Stoica, A.: Evolvable hardware for autonomous systems. In: CEC Tutorial (2004)

Strojnik, P., Peckham, P.H.: Implantable Stimulators for Neuromuscular Control. In: Bronzino, J.D. (ed.) The Biomedical Engineering Handbook, 2nd edn. CRC Press LLC, Boca Raton (2000)

Strohmenger, K., Kakerow, R., Henkel, F., Köther, D.: Re-configurable Multi-Mode Radio Architectures for enhanced 3G Terminals. In: 10th Meeting of Wireless World Research Forum (WWRF) WG6 Reconfigurability, New York, October 27-28 (2003)

Takahashi, E., Murakawa, M., Kasai, Y., Higuchi, T.: Power dissipation reductions with genetic algorithms. In: Proc. of NASA/Dod Conference 2003, pp. 111–116 (2003)

Tarim, T.B., Ismail, M.: Statistical Design of a 10 bit Current-Division Network. Analog Integrated Circuits and Signal Processing, 221–229 (2001)

Thompson, A.: Silicon Evolution. In: Proceedings of the First Annual Conference on Genetic Programming, pp. 444–452. MIT Press, Cambridge (1996)

Tianfield, H.: A Study on the Multi-agent Approach to Large Complex Systems. In: Palade, V., Howlett, J.R., Jain, L. (eds.) Knowledge-Based Intelligent Information and Engineering Systems, Part I, pp. 438–444. Springer, Heidelberg (2003)

Tomassini, M., Sipper, M.: An Introduction to Evolvable Hardware. EvoNews (3), 8–10 (1997)

Torresen, J.: A Scalable Approach to Evolvable Hardware. Genetic Programming and Evolvable Machines 3(3), 259–282 (2002)

Torresen, J.: Increased complexity evolution applied to evolvable hardware. In: Proceedings of ANNIE 1999. ASME Press (1999)

Torresen, J.: Evolvable Hardware – The Coming Hardware Design Method? In: Kasabov, N. (ed.) Neuro-fuzzy Tools and Techniques. Springer, Heidelberg (1997)

Torresen, J.: A divide-and-conquer approach to evolvable hardware. In: Sipper, M., Mange, D., Pérez-Uribe, A. (eds.) ICES 1998. LNCS, vol. 1478, pp. 57–65. Springer, Heidelberg (1998)

Torresen, J.: An evolvable hardware tutorial. In: Becker, J., Platzner, M., Vernalde, S. (eds.) FPL 2004. LNCS, vol. 3203, pp. 821–830. Springer, Heidelberg (2004)

Tuttlebee, W. (ed.): Software Designed Radio.Enabling Technologies. John Wiley & Sons Ltd., Chichester (2002)

Tuttlebee, W. (ed.): Software Designed Radio. Origins, Drivers and International Perspectives. John Wiley & Sons Ltd., Chichester (2002)

Upegui, A., Thoma, Y., Sanchez, E.: The Perplexus bio-inspired reconfigurable circuit. In: Arslan, T., Stoica, A., et al. (eds.) Proceedings of 2007 NASA/ESA Conference on Adaptive Hardware and Systems, Edinburgh, UK, August 5-8, pp. 600–605 (2007)

Vogler, J., et al.: Equipment Management and Control Architecture, E2R white paper (July 2005)

Vassilev, V.K., Job, D., Miller, J.F.: Towards the Automatic Design of More Efficient Digital Circuits. In: EH 2000: Proceedings of the 2nd NASA/DoD workshop on Evolvable Hardware, pp. 151–160. IEEE Computer Society, Los Alamitos (2000)

Vassilev, V.K., Miller, J.F.: Scalability Problems of Digital Circuits Evolution. Evolvability and Efficient Designs. In: EH 2000: Proceedings of the 2nd NASA/DoD workshop on Evolvable Hardware, pp. 55–64. IEEE Computer Society, Los Alamitos (2000)

Venkata, N.D., Evans, B.L.: An Automated Framework for Multicriteria Optimization of Analog Filter Design. IEEE Trans. on Circuits and System-II: Analog and Digital Signal Processing 46(8), 981–990 (1999)

*** (Virtual Computing Company), Virtual Computing – The Virtual Computer is a Reconfigurable Hardware System. EvoNews (3), 10–13 (March 1997)

Vizcaino, A., Favela, J., Piattini, M.: A Multi-agent System for Knowledge Management in Software Maintemance. In: Palade, V., Howlett, J.R., Jain, L. (eds.) Knowledge-Based Intelligent Information and Engineering Systems, Part I, pp. 413–421. Springer, Heidelberg (2003)

Wang, C.-S., Huang, P.-C.: A CMOS Low-IF Programmable Gain Amplifier with Speed – Enhanced DC Offset Cancellation, Rochester NY, pp. 133–136 (2002)

Wang, J., Gulari, M., Bhatti, P.T., Arcand, B.Y., Friedrich, C.R., Wise, K.D.: A Cochlear Electrode Array with Built-in Position Sensing. In: 18th IEEE International Conference on Micro Electro Mechanical Systems (MEMS 2005), Miami (January 2005)

Wang, J., Gulari, M., Wise, K.D.: An Iintegrated Position-Sensing System for a MEMS-Based Cochlear Implant. In: 18th IEEE International Conference on Micro Electro Mechanical Systems (MEMS 2005), Miami (January 2005)

Wakerly, J.F.: Digital Design: Principles and Practices. Prentice-Hall, Inc., Upper Saddle River (2000)

Wasiewicz, P., Mulawka, J.J.: Molecular genetic programming. In: Soft Computing, vol. 5, pp. 106–113. Springer, Heidelberg (2001)

Watanabe, Y., Sato, S., Ishida, Y.: An Approach for Self-repair in Distributed System Using Immunity-Based Diagnostic Mobile Agents. In: Negoita, M.G., Howlett, R.J., Jain, L.C. (eds.) KES 2004. LNCS, vol. 3214, pp. 504–510. Springer, Heidelberg (2004)

Weigel, R., Maurer, L., Pimingsdorfer, D., Springer, A.: RF Transceiver Architectures for W-CDMA Systems Like UMTS: State of the Art and Future Trends, Invited Paper (2001)

Wu, J.-T.: Analog Integrated Circuits. Lectures Notes, National Chiao-Tung University, Taiwan (2003)

Wu, L., Basaran, U., Tao, R., Berroth, M., Boos, Z.: A 2 GHz CMOS dB-Linear Programmable-Gain Amplifier with 51 dB Dynamic range. In: 13th GAAS Symposium, Paris, pp. 617–620 (2005)

*** (Xilinx), The programmable logic data book, Xilinx, San Jose (1995)

yao, X., Higuchi, T.: Promises and Challenges of Evolvable Hardware. IEEE Tran. on Systems, Man, and Cybernetics- Part C: Applications and Reviews 29(1), 87–97 (1999)

Yang, X., Furuhashi, T., Obata, K., Uchikawa, Y.: Constructing a High Performance Signature Verification System Using A GA Method. In: Proceedings of ANNES 1995, Dunedin, New Zealand, pp. 170–173 (1995)

Yoo, J., Hajela, P.: Immune Network Simulations in Multicriterion Design. Structural Optimization 18(2/3), 85–94 (1999)

Zadeh, L.A., Nikravesh, M.: Perception-Based Intelligent Decision Systems. ONR Summer 2002, Program Review, Computer Science Division Department of EECS UC Berkeley (2002), http://www-bisc.cs.berkeley.edu

Zadeh, L.A.: From Search Engines to Question-Answering Systems – The Need For New Tools, Computer Science Division Department of EECS UC Berkeley (2003), http://www-bisc.cs.berkeley.edu

Zebulum, R.S., Mojaradi, M., Stoica, A., Keymuelen, D., Daud, T.: Self-Reconfigurable Analog Arrays: Off-The Shelf Adaptive Electronics for Space applications. In: Proceedings of 2007 NASA/ESA Conference on Adaptive Hardware and Systems, Edinburgh, UK, August 5-8, pp. 529–536 (2007)

Zebulum, R.S., Keymeulen, D., Duong, V., Guo, X., Ferguson, M.I., Stoica, A.: Experimental Results in Evolutionary Fault-Recovery for Field Programmable Analog Devices. In: Proceedings of The 2003 NASA/DoD Conference on Evolvable Hardware, Chicago, Illinois, July 9-11, pp. 182–188 (2003)

Zebulum, R., Mojarradi, M., Stoica, A., Keymuelen, D., Daud, T.: Self Reconfigurable Analog Arrays: Off-the Shelf Adaptive Electronics for Space Applications. In: NASA/ESA Conference on Adaptive Hardware and Systems, Edinburgh, UK, pp. 529–535 (2007)

Zebulum, R.S., Pacheco, M.A., Vellasco, M.M.: Synthesis of CMOS Operational Amplifiers Through Genetic Algorithms. In: Proceedings of the XI Brazilian Symposium on Integrated Circuit Design, vol. 11(1), pp. 125–128 (1998)

Zebulum, R.S., Pacheco, M.A., Vellasco, M.M.: Evolutionary Electronics: Automatic Design of Electronic Circuits and Systems by Genetic Algorithms. CRC Press, Boca Raton (2001)

Zebulum, R.S., Stoica, A., Keymeulen, D., Ferguson, M.I.: Case Studies on the Evolutionary Synthesis of Computational Circuits and Filters and Complete Fuzzy Systems. In: GECCO Late Breaking Papers 2002. pp. 520–527 (2002)

Zebulum, R.S., Pacheco, M.A., Vellasco, M.M.: Evolvable systems in hardware design taxonomy, survey and applications. In: Evolvable Systems: From Biology to Hardware (ICES 1996), pp. 344–358 (1996)

Zhuang, N., Haomin, W.: A new design of the CMOS full adder. IEEE journal of solid-state circuits 7(5), 840–844 (1992)

Zierhofer, C., Hochmair, E.: High-efficiency coupling- insensitive transcutaneous power and data transmission via an inductive link. IEEE Trans. Biomed. Eng. 37(7), 716–722 (1990)

Zierhofer, C., Hochmair, E.: Geometric approach for coupling enhancement of magnetically coupled coils. IEEE Trans. Biomed. Eng. 43(7), 708–714 (1996)

Zierhofer, C., Hochmair-Desoyer, I., Hochmair, E.: Electronic design of a cochlear implant for multichannel high-rate pulsatile stimulation strategies. IEEE Trans. on Rehab. Eng. (1), 112–116 (1995)

Zimmermann, R., Gupta, R.: Low-power logic styles: Cmosvscpl. In: Proceedings of the 22nd European Solid-State Circuits Conference, Neuchatel, Switzerland, pp. 112–115 (1996)

Zrenner, E.: Will Retinal Implants Restore Vision? Science 295(5557), 1022–1025 (2002)

The manufacturer's authorised representative in the EU is Springer Nature Customer Service Centre GmbH, Europaplatz 3, 69115 Heidelberg, Germany. If you have any concerns regarding our products, please contact ProductSafety@springernature.com

Printed and bound by CPI Group (UK) Ltd, Croydon, CR0 4YY

29/07/2026

02180593-0001